The Invention of the Electromotive Engine

B. J. G. van der Kooij

This case study is part of the research work being completed in preparation for a doctorate-dissertation to be submitted to the University of Technology, Delft, The Netherlands (www.tudelft.nl). It is part of a series of case studies about "Innovation" under the title "The Invention Series."

About the text—This is a scholarly case study describing the historic developments that resulted in electromotive engines. It is based on a large number of historic and contemporary sources. As we did not conduct any research into primary sources, we made use of the efforts of numerous others by citing them quite extensively to preserve the original character of their contributions. Where possible we identified the individual authors of the citations. When an author is not identifiable, we identified the source of the text. Facts that are considered to be of a general character in the public domain are not cited.

About the pictures—Many of the pictures used in this study were found on websites accessed through the Internet. Where possible they were traced to their origins, and the source is indicated. As most photos are past the age where copyright would apply, we feel that we make fair use of the pictures to illustrate the scholarly case, and this use is not an infringement of copyright.

Copyright © 2015 B. J. G. van der Kooij

Cover art is line drawing of Bush' electric dynamo (US Patent № 189.997) and Dobrowolsky's electric motor (US patent № 469.515).
(courtesy USPTO)

Version 1.1 (March 2015)

All rights reserved.

ISBN: 1503095878
ISBN-13: 978-1503095878

Contents

Contents i
Preface v
   About the Invention Series vii
   About our research viii
   About this case study x
Context for the discoveries 1
   The nineteenth century 3
      A time of changes: political, economic, social, technical 3
   Science discovers and applies electricity 24
      Electricity as phenomenon: the nature of lightning 32
      Electricity explored 38
      Creation of electromagnetic power 63
      Electrochemistry: the wet cell 66
   The power of lightning understood 69
The invention of the electric DC motor 71
   The direct current electric motor 71
      The electromagnetic reciprocal engine 72
      The electromagnetic rotatory engine 76
      The invention of the DC motor 82
      A cluster of innovations for the DC motor 85

The invention of the electric dynamo ... 87

Electricity generators: the dry battery ... 88
- Principle of the dynamo: DC/AC ... 90
- Early magneto-electric dynamos ... 91
- The self-exciting electric dynamo ... 108
- Some of the later dynamos ... 123
- From Component to System ... 125
- The invention of the magneto-electric dynamo ... 128
- A cluster of innovation for the dynamo ... 129

The Electric Revolution ... 135
- Industrial bonanza: cluster of businesses ... 135
- Booming markets ... 140
- Early electric systems ... 142

The era of power ... 143
- Mobile applications of DC motors ... 144
- Stationary applications of DC motors ... 153

The invention of the electric induction motor ... 159

Alternating current: electromagnetic components ... 162
- From induction coil to three-phase transformer ... 162

Alternating current: the induction motor ... 166
- Science and engineering discover induction ... 166
- Early induction motors ... 169
- The polyphase induction motor ... 172

Parallel development of the AC generator ... 184
- Later versions of induction motors ... 190
- The invention of the AC induction motor ... 193
- A cluster of innovation for the AC induction motor ... 197

The development of the electric alternating-current power system ... 199
- AC-distribution networks ... 200
- George Westinghouse (1846–1914) ... 206

Battle of currents: DC versus AC ... 217
- Edison versus Westinghouse ... 218

Conclusion 227
   Human curiosity, ingenuity, and competition 228
      Curiosity in the nature of lightning 228
      Ingenuity 230
      Competition 231
   Societal change induced by technical change 232
      Second Industrial Revolution: "Power to the people" 232

References 239

About the author 253

B.J.G. van der Kooij

Preface

When everything is said and done,
and all our breath is gone,
The only thing that stays
Is history, to guide our future ways.

My lifelong intellectual fascination with technical innovation within the context of society started in Delft, the Netherlands, in the 1970s at the University of Technology—at both the electrical engineering school and the business school.[1] Having been educated as a technical student with vacuum tubes, followed by transistors, I found the change and novelty of the new technology of microelectronics to be mind-boggling—not so much from a technical point of view but because of all the opportunities for new products, new markets, and new organizations, with a potent technology as the driving force.

During my studies at both the School of Electric Engineering and the School of Business Administration,[2] I was lucky enough to spend some time in Japan and California, where I observed how cultures influence the context for technology-induced change and what is considered novel. In Japan I explored the research environment; in Silicon Valley I saw the business environment—from the nuances of the human interaction of the Japanese, to the stimulating and raw capitalism of the United States. The technology forecast of my engineering thesis made the coming technology push a little clearer: the personal computer was on the horizon. The implementation of innovation in small and medium enterprises, the subject of my management thesis, left me with a lot of questions. Could something like a Digital Delta be created in the Netherlands?

During the journey of my life, innovation has been the theme. For example, in the mid-1970s, I joined a mature electric company that

---

[1] At present the schools referred to are called the Electrical Engineering School at the Delft University of Technology and the School of International Business Administration at the Erasmus University Rotterdam.
[2] The institutions' actual names were Afdeling Electro-techniek, Vakgroep Mikro-Electronica, and Interfaculteit Bedrijfskunde.

manufactured electric motors, transformers, and switching equipment, and business development was one of my major responsibilities. How could we change an aging corporation by picking up new business opportunities? Japan and California were again on the agenda but now from the business point of view. I explored acquisition, cooperation, and subcontracting. Could we create business activity in personal computers?

The answer was no.

I entered politics and became a member of the Dutch Parliament (a quite innovative move for an engineer), and innovation on the national level became my theme. How could we prepare a society by creating new firms and industries to meet the new challenges that were coming and that would threaten the existing industrial base? What innovation policies could be applied? In the early 1980s, my introduction of the first personal computer in Parliament caused me to be known as "Mr. Innovation" within the small world of my fellow parliamentarians. Could we, as politicians, change Dutch society by picking up the new opportunities technology was offering?

The answer was no.

The next phase on my journey brought me in touch with two extremes. A professorship in the Management of Innovation at the University of Technology in Eindhoven gave me room for my scholarly interests. I was looking at innovation at the macro level of science (part time). I started a venture company making application software for personal computers, and that satisfied my entrepreneurial obsession. Now I began to concentrate (nearly full time) on the implementation of innovation on the microscale of a starting company. With my head in the scientific clouds and my feet in the organizational mud, I was stretching my capabilities. At the end of the 1980s, I had to choose, and entrepreneurship won for the next eighteen years. Could I start and do something innovative with personal computers myself?

The answer was yes.

When I reached retirement in the 2010s and reflected on my past experiences and the changes in our world since those 1970s, I wondered what made all this happen. Technological innovation was the phenomenon that had fascinated me along my entire life journey. What is the thing we call "innovation"? In many phases of the journey of my life, I tried to formulate an answer: starting with my first book, *Micro-computers, Innovation in Electronics* (1977, technology level), next with my second book, *The Management of Innovation* (1983, business level), and my third book, *Innovation, from Distress to Guts* (1988, society level). In the 2010s I had time on my hands. So I decided to pick up where I left off and start studying the subject of innovation again. As a guest of my alma mater, working on my dissertation, I tried to find an answer to the question "What is innovation?"

It started in Delft. And, seen from an intellectual point of view, it will end in Delft.

B. J. G. van der Kooij

About the Invention Series

Our research into the phenomenon of innovation, focusing on technological innovation, covered quite a timespan: from the late seventeenth century up to today. The case study of the steam engine marks the beginning of the series. That is not to say there was no technological innovation before that time. On the contrary, *imitation, invention,* and *innovation* have been with us over a much longer time. But we had to limit ourselves, as we wanted to look at those technological innovations that were the result of a general-purpose technology (GPT). Clearly some clarification is needed here, so we will define the major elements of our research: innovation, technology, and GPT.

We define *innovation* as the creation of something new and applicable. It is a process over time that results in a new artifact, a new service, a new structure, or a new method. Where invention is the discovery of a new phenomenon that does not need a practical implementation, innovation brings the initial idea to the marketplace where it can be used. We follow Alois Schumpeter's differentiation: "Innovation combines factors in a new way, or that it consists in carrying out New Combinations" (Schumpeter, 1939, p. 84). Innovation is quite different from invention: "Although most innovations can be traced to some conquest in the realm of either theoretical or practical knowledge, there are many which cannot. Innovation is possible without anything we should identify as invention, and invention does not necessarily induce innovation, but produces of itself…no economically relevant effect at all" (Schumpeter, 1939, p. 80). What about invention then? We follow here Abott Usher's interpretation where the creative act is the new combination of the "Act of skills" and the "Act of insight": "Invention finds its distinctive feature in the constructive assimilation of preexisting elements into new syntheses, new patterns, or new configurations of behavior" (Usher, 1929, p. 11).

We define *technology* as the know-how (knowledge) and way (skill) of making things. Technology is more than "technique," which is where it originates from. "Technology is a recent human achievement that flourished conceptually in the eighteenth century, when technique was not more seen as skilled handwork, but has turned as the object of systematic human knowledge and a new 'Weltanschauung' (at that time purely mechanistic)" (Devezas, 2005, p. 1145). We follow Anne Bergek, et al., here: "The concept of technology incorporates (at least) two interrelated meanings. First, technology refers to material and immaterial objects—both hardware (e.g., products, tools, and machines) and software (e.g., procedures/processes and digital protocols)—that can be used to solve real-world technical problems. Second, it refers to technical knowledge, either in general terms or in terms of knowledge embodied in the physical artifact" (Bergek, Jacobsson, Carlsson, Lindmark, & Rickne, 2008, p. 407).

We define a *general-purpose technology* as the cluster of technologies that result in innovations that have considerable impact on society: "…the pervasive technologies that occasionally transform a society's entire set of economic, social, and political structures" (Lipsey, Carlaw, & Bekar, 2005, p. 3). GPT results in what we are identifying as the Industrial Revolution and the Information Revolution. It is the engine of economic growth, but it is also the engine of technical, social, and political change—the engine of creative destruction. We follow the definition of Richard Lipsey, et al.: "A GPT is a technology that initially has much scope for improvement and eventually to be widely used, to have many uses, and to have many spillover effects" (*Ibidem*, p. 133). The GPT is not a single-moment phenomenon; it develops over time: "They often start off as something we would never call a GPT (e.g., Papin's steam engine) and develop in something that transforms an entire economy (e.g., Trevithick's high-pressure steam engine)" (*Ibidem*, p.97).

The case studies are about observing phenomena as they occur in the real world—for example the development of the steam engine (from which one can conclude it was a GPT according to the definition). The observation of what caused the Second Industrial Revolution is more complex. Is electricity the GPT, or are the electric motor and the electric dynamo the GPT? Or can it be that the resulting development of the electric light and telegraph is a GPT on its own? The interpretation becomes more complex and the opinions diffused—especially when one looks at the present time, for example the phenomena of the Internet.

About our research

This book is the second manuscript in the Invention Series—a series of books about inventions that created the world we live in today. In the first manuscript, *The Invention of the Steam Engine*, we explored a methodology to observe and investigate the complex phenomena of "technological innovation" as part of a General Purpose Technology (GPT). In that case it was about the *steam technology* that fueled the Industrial Revolution. One could consider that case study as a trial to determine if our methodology could be accurately applied. The result was promising enough to try again. So let's describe the basic elements of our research.

In this case study, our *field of interest* is the GPT of electricity. To understand how this technology could fuel the next Industrial Revolution, we applied the method of the case study. The case-study method offers room for "context and content." The *context* is the real-life context: the scientific, social, economic, and political environment in which the observed phenomena occurred. The *content* is the technical, economic, and human details of those phenomena. The reader will recognize these aspects in the structure of the manuscript.

The case study is based on a specific *scholarly view*: to observe the phenomena as they occurred in the real world. This view is based on the construct of "clusters of innovations" as identified by early twentieth century scholars active in the Domain of Innovation Research. Foremost among those economists was Alois Schumpeter, who related the clusters of innovations to business cycles under the influence of creative destruction: "…because the new combinations are not, as one would expect according to general principles of probability, evenly distributed through time…but appear, if at all, discontinuously in groups or swarms" (Schumpeter & Opie, 1934, p. 223). "The business cycle is a direct consequence of the appearance of innovations" (*Ibidem*, pp. 227–230). For Schumpeter it was the entrepreneurs who realized the innovation and, as imitators were soon following in the entrepreneurial act, were thus creating the "clusters of business". Clusters that are nested within business cycles and the resulting economic waves. Later it was Gerhard Mensch and Jaap van Duijn who related the basic innovation within the clusters to the long waves in economy, respective to industrial cycles. Mensch related the cyclic economic pattern to basic innovations: "The changing tides, the ebb and flow of the stream of basic innovations explain economic change, that is, the difference in growth and stagnation periods" (Mensch, 1979, p. 135). Van Duijn referred to innovation cycles (Duijn, 1983). More recently it was scholars like Dosi, Tushman, Anderson, and O'Reilly who developed, as part of their view on technological revolutions and technological trajectories, the construct of the dominant design. This Dominant Design we considered to be the basic innovation.

So our *unit of analysis* is the cluster around the basic innovation—including the preceding and derived innovations. We choose for embedded, multiple-case design of the GPT "steam technology" (a collection of many mechanical, hydraulic, and related technologies explored in the first manuscript) and the GPT "electric technology" (idem, this manuscript). The method is *multiple* because we looked simultaneously at the scientific, technical, economic, and human aspects. It is *embedded* because we looked simultaneously at the individuals (the inventors, the entrepreneurs), the organizations (their companies), and societies—thus making the analysis multilevel and multidimensional. Our qualitative data originate from general, autobiographic, and scholarly literature (see References), creating a mix of sources that are quoted extensively. Our quantitative data were sampled from primary sources like the United States Patent and Trademark Office (USPTO).

Our *perspective* was the identification of patterns that are related to the cluster-concept. Can clusters of innovation within a specific General Purpose Technology be identified? If so, how are they related, and how are the clusters put together? The first pilot case showed that it could be done. So in this case study, our objective was to identify the basic innovations that played a dominant role in the GPT of electricity that created the (second) Industrial Revolution. We used *patents* as innovation-identifiers, and we used *patent wars* (patent infringement and patent litigation) and *economic booms* (business creation and business-and-industry cycles) to identify basic innovations. So these aspects are quite dominant in the study. In the scheme below, the "Cluster of Innovations" and the related "Cluster of Businesses" concept is visually represented.

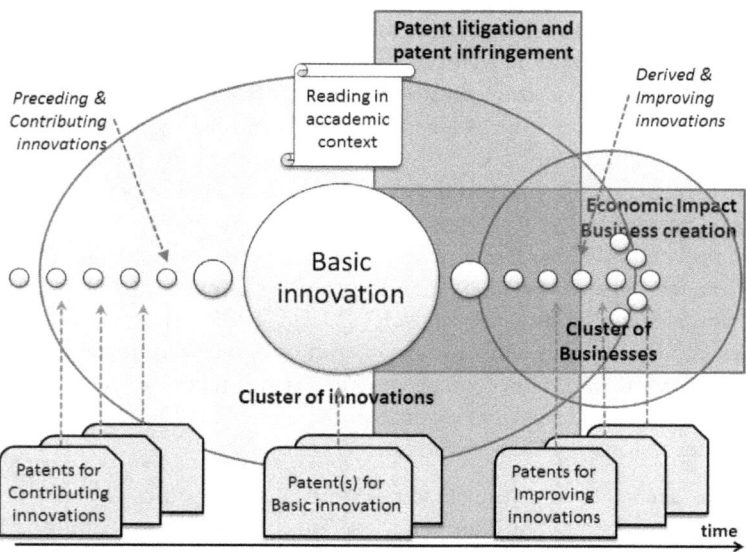

Scheme 1: The construct of the Cluster of Innovations and Cluster of Businesses

About this case study

This case study is the result of our quest in the Nature of Innovation. It is divided in the following sections:

Context for the discoveries: We will begin with a thorough look at the events that created the historical climate of the time. Although these events are not directly related to the invention of electricity, the social, economic and political turmoil—followed by relative peace—created the context for scientific discovery. As we have to limit ourselves we focus on the European theatre where we examine the history of the early nineteenth century. We will describe the early efforts where curious people started to try and understand the 'nature of lightning', just as they had explored the 'nature of heat' before[3].

The invention of the electric DC motor: This segment is about the early form of electricity; the 'direct current' (DC) generated by the wet cell. Here will describe those early efforts that resulted in the creation of the first rudimentary artifacts that used electricity: the electric motor that was supplied by electricity from the Voltaic Battery.

The invention of the electric dynamo: We proceed than with the development of a radical new artifact; the electromagnetic engine that created electricity in abundance, freeing us from the cumbersome batteries of that time. It would create an Electric Revolution with its industrial bonanza that created the Era of Power and the Era of Light[4].

The Invention of the electric AC induction motor: This segment is about de 'alternating current (AC).' Here we pay attention to the distribution of the electricity generated by the electric dynamo's. Finally we look at the artifact that dominated the progress in the application of electricity; the induction motor used in all those industrial and household applications.

This is a story about the *General Purpose Technology of 'electricity'* with its "clusters of innovations" and "clusters of businesses" that changed the world we live in.

---

[3] See: B.J.G.van der Kooij: *The Invention of the Steam Engine* (2015)
[4] See: B.J.G.van der Kooij: *The Invention of Electric Light* (2015)

B.J.G. van der Kooij

The Invention of the Electromotive Engine

B.J.G. van der Kooij

Context for the discoveries

Take a moment in time quite some years ago. For a person in the preelectric era, electricity was a miracle. People were used to manual labour, at home and at work. They were living and working with a dependence primarily on daylight, in poorly lit houses. A situation that would change with the discovery of electricity. It would eliminate a large number of daily household chores, like drawing and hauling water, washing and wringing clothes by hand, and ironing them with a stove-heated flatiron. It would replace the smelly oil lamps, creating safe light. On farms electricity did away with primitive methods of threshing grains, the need to hand milk cows, and the necessity of manually chopping or grinding and pitchforking hay into lofts. In workshops and factories, electricity would replace poorly gas-lit premises—first by incandescent and later by fluorescent lights— while convenient, efficient, and precisely adjustable electric motors would do away with the dangerous transmission belts that were driven by steam engines. Noisy and polluting horse-powered streetcars would be replaced by ones that were powered by electricity. Electricity also changed railways when electricity supplanted inefficient, polluting steam engines (Smil, 2005, p. 35).

Take another—more recent— moment in time. For a person in the twenty-first century, electricity is nothing special—something invisible and taken for granted. Food is routinely stored in cool or deep-freezing facilities, and lights and heating are a standard convenience. Machines are used every day to take over the domestic chores of cleaning and washing, and robot-like production machines are utilized in factories in automated production processes. No business activity seems possible without computers. Visual and auditory communications (television, radio, telephones) are available in abundance. Transportation systems include electrically powered cars, trains, and subways. All these are based on electrical power that is always available. Only when the electrical power system fails, as is often the case after a thunderstorm in the quite populated South of France, does one realize our total dependency on electricity as all private, professional, and business activity comes to a grinding halt.

Society changed substantially between those two moments in time (before and after electrical inventions) due to technical changes initiated by the new phenomenon of electricity.

> *Of the great construction projects of the last century, none has been more impressive in its technical, economic, and scientific aspects, none has been more influential in its social effects, and none has engaged more thoroughly our constructive instincts and capabilities than the electric power system…Electric power systems embody the physical, intellectual, and symbolic resources of the society that constructs them…In a sense electric power systems, like so much other technology, are both causes and effects of social change* (Hughes, 1993, pp. 1-2).

In hindsight the enormous impact of the introduction of electricity in society is clear. But it took quite some time, many scientific discoveries, and a lot of engineering efforts before this all came to happen within the context that existed in the nineteenth century.

Scientific curiosity had already started to play an important role in the period before 1800. Much happened that created the foundations for later developments. However, as this is not the place to discuss science's general development over time (as in the "history of science"), we will limit ourselves to the context for the explorations into the phenomenon of electricity—or the "power of lightning" as it was called in those days. We will look at the context for those scientific developments that created the "electric technologies." Developments that were based on the observations and experiments of many scientist and engineers—thinkers and tinkerers—trying to create an understanding of the nature of the phenomenon at hand.

This context defines the developments that resulted in the invention of electromotive engines.[5] The context is European and American. The story is about the madness of the time and the creativity and perseverance of individuals.

---

[5] The content of this part of the case is not the result of my own *primary* research, but is based on other scholarly work. I drew information from a broad range of sources, including Wikipedia and sources found through Google Scholar. Where realistically possible these sources are acknowledged.

The nineteenth century

Technical changes have to be seen in the context of their specific time. Certainly when *technical change* is leading to social, economic and political change. as was the case with the Industrial Revolutions. But it is also the other way around where social, economic and political change sets the context for technical change. The same goes for *economic change*. New technological developments create economic changes when new industries emerge, creating new jobs, organizational forms and organizational structures. And, in its turn, prospering economies create a favorable context for technical renewal. As also the advent of 'electricity', the focus of this case study, has to be seen in the context of its time, we will try and describe some major aspects of a time where so much changed over such a relative short period of time.

A time of changes: political, economic, social, technical

Europe at the end of the eighteenth century and beginning of the nineteenth century had seen many wars and national uprisings. The French Revolution of 1789 was followed by the French Revolutionary Wars (1792–1802) and the French occupancy of the Low Countries (Netherlands and Belgium), the Rhinelands (west bank of the Rhine), northern Italy, parts of Spain, Switzerland, and the Savoy and Liguria regions. These wars were followed by the Napoleonic Wars (1803–1815) and the rise of the First French Empire (Figure 1). A period that started with the Battle of Austerlitz (in the Czech Republic) ended with collapse after the disastrous invasion of Russia in 1812 and the battle of Waterloo in Belgium (1815) (Figure 2).

The Battle of Waterloo was fought on Sunday, June 18, 1815, near the hamlet of Waterloo in present-day Belgium—then part of the United Kingdom of the Netherlands. An Imperial French army under the command of Emperor Napoleon was defeated by the armies of the Seventh Coalition, comprised of an Anglo-allied army under the command of the Duke of Wellington combined with a Prussian army under the command of Gebhard von Blücher. It was the culminating battle of the Waterloo Campaign, and it was Napoleon's last war.

Figure 1: The First French Empire, 1804–1814

The French Empire is the darkest area, while the "Grand Empire" includes areas under French military control (lighter grey) and the allies of France.

Source: Wikimedia Commons

Wellington's troops consisted of 67,000 men: 50,000 infantry, 11,000 cavalry, and 6,000 artillery with 150 guns. Of these, 25,000 were British, with another 6,000 from the King's German Legion. In addition, there were 17,000 Dutch and Belgian troops—11,000 from Hanover, 6,000 from Brunswick, and 3,000 from Nassau. When the battle was over, Waterloo had cost Wellington around 15,000 dead or wounded and Blücher some 7,000. Napoleon lost 25,000 dead or wounded, with 8,000 taken prisoner.

Waterloo was a landmark, a decisive battle, in more than one sense. It definitively ended the series of wars that had convulsed Europe and had involved many other regions of the world, starting with the French Revolution of the early 1790s. Finally, it ushered in almost half a century of international peace in Europe; no further major conflict occurred until the Crimean War (1853–1856).

Figure 2: Battle of Waterloo, 1815
Source: Wikimedia Commons, Painting by William Sadler

> *The battle of Waterloo had some interesting aspects related to it. Firstly, the major British financier of the war efforts, the banker Nathan Rothschild, got news of the battle by his private intelligence network (agents and couriers whose duty it was to follow in the wake of armies assisted by pigeons). In London, the information then available was that in the beginning of the battle Napoleon's forces seemed to be winning. This situation changed, however, drastically the next day. Rothschild's messenger Rothworth, present on the battlefield of Waterloo, after verifying that Napoleon's forces were defeated, travelled by horse to Ostende where he paid 2.000 Francs to have a sailor transport him—in quite bad weather—to England across the English Channel. Once Nathan Rothschild obtained the delivery of the news on the 20th of June, he used his influence to reconfirm that the battle was lost and began to sell all his English stock, advising the financial world to do the same. Thus, everyone believing Wellington to be*

defeated began selling, causing stocks to plummet to practically nothing. At the last minute, his agents secretly began buying up all the stocks at rock-bottom prices. On June 21ˢᵗ, Wellington's envoy, Major Henry Percy, arrived at the War Office reporting that Napoleon had lost a third of his men in battle. Immediately this news caused stock prices to soar, giving the Rothschild family a healthy million sterling pound profit and complete control of the British economy. (Reeves, 1887, pp. 169-175)

Secondly, as an interesting detail, after the battle the Duke of Wellington was awarded the everlasting title Prince of Waterloo. In addition to this title, he was granted 1.050 hectares of land near the battlefield and a yearly donation of 20.000 guilders.[6] Member Bruno Stevensheydens questioned the issue of the yearly payment in the Belgium Parliament in June 2009. The Belgian Finance Minister, Didier Reynders, answered that the Belgian government had no intention of backing out of its commitment as part of the Treaty of London (1839). Till today (2015) the descendants of the Duke of Wellington[7] are still cashing in from the labors of some seventy farmers who live on the land near Waterloo, to the tune of around £100.000 or US $160,000 a year.[8]

Belgium, then part of the present-day Netherlands, became independent in 1830 with the Belgian Revolution (a situation confirmed in the Treaty of London in 1839). Under the treaty the European powers recognized and guaranteed the independence and neutrality of Belgium.

Jumping forward in time from this landmark in the mid-1810s, about a century after Napoleon's defeat at Waterloo) the Great War (World War I) started in 1914. The great powers of the world, read "Europe," were at war. Now it was the *allied forces* (England, France, and Russia) that battled the

---

[6] Equivalent to about € 145 million in 2010. Source International Institute of Social History, http://www.iisg.nl/hpw/calculate2.php.

[7] For example the eighth Duke of Wellington, Arthur Wellesley, Prince of Wellington, Duke of Vittoria an Earl of Mornington, owns a 7,000-acre Hampshire estate, 20,000 acres of Belgium and Spain. The estate was thought to be worth £50m in 2001. "Ten dukes-a-dining: Gathered together over lunch for a unique picture, the grandees with £2bn and 340,000 acres between them," last modified October 7, 2009, http://www.dailymail.co.uk/news/ article-1218628/Ten-dukes-dining-Gathered-lunch-unique-picture-grandees-2bn-340-000-acres-them.html#ixzz3HnDYfLhF.

[8] Information is based on the following research: "Battle of Waterloo," "Nathan Mayer Rothschild," "Prince of Waterloo," http://www.wikipedia.org/; John Tagliabue, "Still Battling at Waterloo," *International Herald Tribune*, September 25, 2013; 'Battle over legacy of Waterloo," BBC, January 19, 2000, http://news.bbc.co.uk/2/hi/europe/609869.stm; Belgische Kamer van Volksvertegenwoordigers, "Schriftelijke vragen en antwoorden," QRVA5266, 22-06-2009, http://www.dekamer.be/doc/qrva/pdf/52/52k0066.pdf. (accessed January 2015)

central powers (Germany, Austria-Hungary). Europe was in turmoil. The Russian Revolution of February 1917 had overthrown the tsarist autocracy. In defeated Germany the German Revolution of 1918–1919 had replaced the imperial government of Kaiser Wilhelm II and created the Weimar republic (Figure 3). The early decades of the nineteenth century and those of the twentieth century had quite dramatic landmarks.

Figure 3: The Weimar Republic (1919)
Source: Wikimedia Commons

It was the period between these landmarks, Waterloo and World War I, in which Europe had its share of turmoil. This turmoil was caused by social changes, cultural changes, technological changes, and political changes. It was also caused by economic, industrial, and agricultural factors, where former powers (clergy, royalty) were challenged by new powers (rise of the middle classes, liberalism, and socialism). The period after the French Revolution (1789) culminating in the period of the European Revolutions of 1848, saw a transformation of the western world as never seen before. Influenced by the First Industrial Revolution, this period was characterized by the historian Eric Hobsbawn[9] in this way:

> *The great revolution of 1789–1848 was the triumph not of "industry" as such, but of capitalist industry; not of liberty and equality in general but of middle class or "bourgeois" liberal society; not of "the modern economy" or "the modern state," but of the economies and states in a particular geographical region of the world (part of Europe and a few patches of North America), whose center was the neighboring and rival states of Great Britain and France. The transformation of 1789–1848 is essentially the twin upheaval which took place in those two countries, and was propagated thence across the entire world…The historic period*

---

[9] For the records, a personal note: The Jewish British Marxist historian Eric Hobsbawm (1917–2012), writer of the majestic *The Age of Revolution: Europe 1789–1848*, *The Age of Capital: 1848–1875*, and *The Age of Empire: 1875–1914* was for decades under surveillance by the MI5, Britain's domestic security service. As a member of different socialist/communist organizations (such as the Socialist History Society), he and his college historian, Christopher Hill, had their phones tapped, correspondence intercepted, and their friend and wives monitored. "Hobsbawm, who became one of Britain's most respected historians and was made a Companion of Honour while Tony Blair was prime minister, first came to the notice of MI5 in 1942 when he and thirty-eight colleagues were described as being 'obvious members of the CPGB [the Communist Party of Great Britain] on Merseyside.'" Richard Norton-Taylor, *The Guardian*, October 24, 2014.

which begins with the construction of the first factory system of the modern world in Lancashire and the French Revolution of 1789, ends with the construction of its first railway network and the publication of the Communist Manifesto (Hobsbawm, 2010, pp. 1-4).

One could say that the first half of the nineteenth century was shaped by the first Industrial Revolution[10]. It dramatically changed the way societies existed, founded by their histories, in half a century. It was a period that left a legacy from the times in which the French dominated a large part of Europe, even long after Napoleon was defeated at Waterloo:

In all these territories…the institutions of the French Revolution and the Napoleonic Empire were automatically applied, or were the obvious models for local administration: feudalism was formally abolished, French legal codes applied, and so on. These changes proved far less reversible than the shifting of frontiers. Thus the Civil Code of Napoleon remained, or became once again, the foundation of local law in Belgium, in the Rhineland (even after its return to Prussia), and in Italy. Feudalism, once officially abolished, was nowhere re-established…But changes in frontiers, laws, and government institutions were as nothing compared to a third effect of these decades of revolutionary war: the profound transformation of the political atmosphere…It was now known that revolution in a single country could be a European phenomenon; that its doctrines could spread across the frontiers and, what was worse, its crusading armies could blow away the political systems of a continent. It was now known that social revolution was possible; that nations existed as something independent of states, peoples as something independent of their rulers, and even that the poor existed as something independent of the ruling classes. (Hobsbawm, 2010, pp. 90-91).

The European Revolutions of 1848

The *European Revolutions* of 1848 were a series of upheavals throughout Europe that resulted in significant social and cultural changes (Figure 4).[11] The revolutionary wave began in France in February 1848 and immediately spread to most of Europe. Quite a few factors were involved. These factors included the widespread dissatisfaction with political leadership; demands for more participation in government and democracy; the demands of the working classes; the upsurge of nationalism; and finally, the regrouping of the reactionary forces based on the royalty, the aristocracy, the army, and the peasants.

---

[10] For details see:: B.J.G. van der Kooij, *The Invention of the Steam Engine.* (2015)
[11] See also: (Taylor, 2000), (Robertson, 1952), (R. J. W. Evans, 2000).

In the *German states* (Germany was a collection of small states, such as the country we call Italy today), revolution was caused by the popular desire for increased political freedom, liberal state policies, democracy, nationalism, and for freedom from censorship. Members of the middle class were committed to liberal principles while the working class sought radical improvements in their working and living conditions. Disastrous economic conditions also played a part. A cholera epidemic led to widespread death and suffering in Silesia. Population growth and the failures of harvests in 1846 and 1847 caused famine and misery. Many people moved to the cities in order to survive, but wages were very low, and living conditions were appalling.

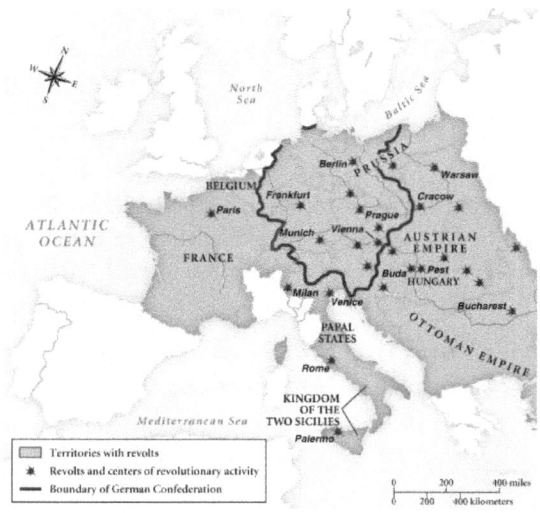

Figure 4: European countries with revolutions in 1848

Source: http://www.age-of-the-sage.org

In *Italy* the revolt was against the Bourbon rule over the northwest part of Italy, the Austrian control over northern Italia, and the papal control in central Italy. In Sicily the people began to demand a provisional government, separate from the government of the mainland. These revolts in Sicily helped to spark revolts in the northern Kingdom of Lombardy—Venetia. Revolutions in the Lombardy city of Milan forced about twenty thousand of Austrian General Radetsky's troops to withdraw from the city.

Figure 5: French February Revolution of 1848

Lamartine in front of the Town Hall of Paris rejects the red flag on February 25, 1848.

Source: Wikimedia Commons, Henri Félix Emmanuel Philippoteaux

In *France* the revolution was driven by nationalist and republican idealists among the French general public, who believed the people should rule themselves. It was spurred on by a financial crisis and bad harvests in 1848, and the following year saw an economic depression. Many of the participants in the 1848 French Revolution were of the so-called "petite bourgeoisie" (the owners of small properties, merchants, shopkeepers, etc.). Indeed the "petite" or "petty bourgeoisie" outnumbered the working classes (unskilled laborers working in mines, factories, and stores, paid for their ability to perform manual labor and other work rather than for their expertise) by about two-to-one in 1848. However, the financial position of the petty bourgeoisie was extremely tenuous.

In the *Austrian Empire*, much of the revolutionary activity was of a nationalist character. Besides these nationalisms, liberal and even socialist currents resisted the empire's long-standing conservatism. In *Poland* the Greater Poland Uprising of 1848 was an unsuccessful military insurrection of Poles against Prussian forces. It was a reaction to the German colonization that had grown in strength, and when policies against Polish religion and traditions were introduced, the local population begun to feel hostile towards the Prussian and German presence. In *Denmark* the growing bourgeoisie had demanded a share in government, and in an attempt to avert the sort of bloody revolution occurring elsewhere in Europe, Frederick VII gave in to the demands of the citizens. In countries such as Great Britain, the Netherlands, Switzerland, Portugal, and Spain the transitions, if any, were more moderate.[12]

Figure 6: Galician slaughter depicting the massacre of Polish nobles by Polish peasants in Galicia in 1846

Source: Wikimedia Commons, Painting by Jan Lewicki

The situation in the *Netherlands* was an interesting development. Here, the Constitution was changed in 1848, and the change stripped power from Willem (Frederick George Lodewijk), Prins van Oranje-Nassau (1792–1849). King Willem II had been the royal ruler since 1840, and the revision

---

[12] See "Revolutions of 1848," www.wikipedia.org (accessed January 2015). See also: (R. J. W. Evans, 2000).

created a parliamentary democracy with the ministerial responsibility we have today. This drastic constitutional revision was not the result of a change of royal mind but, instead, had a completely different background. In a summary referencing Willem II's biography,[13] it reads:

> *Willem II was a driven soldier. During many battles, first in Spain (1811–1813) and later in the southern Netherlands (1815), he fought in the front lines against Napoleon. With hindsight, it is a miracle he only was injured at Waterloo (1815). To rescue the House of Orange dynasty, the Oranges tried from 1809 to marry Prince William with Princess Charlotte, the heiress to the English throne. The failure of the commitment was not William's fault. Charlotte used him in the feud with her father and the English politics became divided on what had become from 1813 an engagement to be married between two heirs…Prince William after 1815 has repeatedly tried to divest the French Bourbon King Louis XVIII from the throne. In 1830 he was offered by the moderate South-Dutch rebels the Belgian throne; he refused…*
>
> *On October 7th, 1840 Willem II was crowned King of the Netherlands. The Netherlands were in an economic crisis, aggravated by the long lasting conflict with Belgium. The separation of Belgium State's required a new constitution. Liberals in Parliament wanted a reform in the Constitution. A group of nine liberals in Parliament—the "Negenmannen"—took the initiative for a much more drastic revision of the Constitution, without any success due to objection from conservative members and the crown. That was the political context. But there was more. As the economy and trade further deteriorated, the crops failed due to a potato disease caused by the fungus "Phytophthora infestans" (the Dutch potato famine of 1846–1847 equaled the Irish potato crisis). Food prices rose, the mortality rates increased resulting in famine casualties in the Netherlands of about 126.000 persons. In 1849 a cholera epidemic occurred (Mokyr, 1980, p. 436). It were additional problems contributing to the already declining economy of the Netherlands (Wilson, 1939). Then there were those 1848-revolts all over Europe, aggravated by the Continental Famine…*
>
> *The Continental Famine was caused by poor harvests of potatoes, due to the same late blight, but also of [crop failures of] grain, due to frost, drought, rust, voles, inopportune rains, floods, and hailstorms. The Continental Famine was enhanced by hoarding, speculation, and poor governance. Hunger was followed by infectious*

---

[13] J. Van Zanten, "Koning Willem II." *Boom*. Press Release. Source: http://www.prinsbernhardcultuurfonds.nl/ t1.asp?path=vi1arvf9 (accessed January 2015)

diseases…*The harvest failures of 1845 and 1846 and the resulting famines came on top of rural pauperization and urban discontent, and thus contributed to the revolutions of 1848 on the European* Continent (Zadoks, 2008, p. 5).

It is within this context that in 1847 Willem II instituted a committee for the reform of the Constitution. The former liberal member of Parliament Johan Rudolph Thorbecke (then a professor in Leiden) became president of the committee. He had a great influence on the creation of the new Constitution as he was participating in the reform movement that, in reaction to the crises, wanted a radical change in the administration.

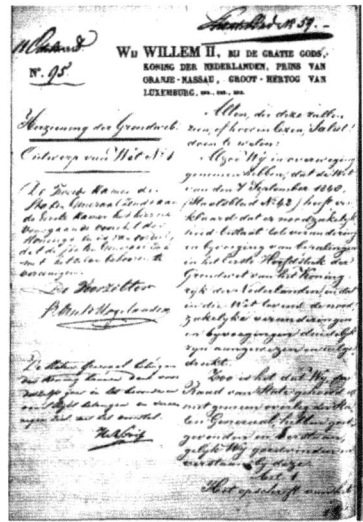

Figure 7: The first page (Preambule) of the 1848 Constitution

Source: http://www.engelfriet.net/ Alie/Aad/willemII.htm

Against the conservative majority, Willem II—surprisingly—agreed with the proposed Constitutional Revision that stripped him of his powers. His decision was made under the pressure of blackmail (the threat of revealing his bisexual relations to the public).[14] It is true that in 1848 his thoughts about constitutional revision were influenced by the revolutions elsewhere in Europe, but, above all, his decisions were guided by a small group of radical journalists and politicians who were aware of his bisexual orientation. They threatened him with disclosure if he was not prepared to make far-reaching political concessions. Largely under pressure from the blackmail of "bastards and schemers," as a minister called them, did Willem II accede in the Constitutional Revision of 1848.

This is an example of the more "moderate" changes in society during the 1848 revolutions. Whatever the causes and specific results in each of the countries, these waves of revolutions in 1848 created a political earthquake that resulted in long-lasting changes.[15]

---

[14] "Koning Willem II gechanteerd wegens homoseksualiteit," *NRC Handelsblad*, November 29, 2013.

[15] The similarity is striking between this era in Europe and the wave of revolutions that swept the Middle East around 2011–2013 and was coined the "Arab Spring." See: (Goldstone, 2011).

Economic changes[16]

At the beginning of the nineteenth century the economies of the European countries were based on landownership by the ruling classes (aristocracy, church) and the servile status of the working class. Most European economies were agriculturally dominated.

> *It may be concluded that except for the Netherlands and the British mainland the major part of the Western European population was rural, living on farms, in villages, or in small cities. About half of the population was dependent directly on agriculture for its living. Clearly, however, a considerable part of the rural population was active outside agriculture, pointing at the fact that even outside large cities specialization was becoming of importance. Rural activities outside agriculture ranged from artisans and salesmen working for the local markets to families working in proto-industry supplementing their income with small agricultural activities mainly directed at self-provision* (Vanhaute, O'Grada, & Paping, 2007, p. 7).

Due to this fact, the economies were also very locally and regionally oriented. The world for common people was relatively small in those days. Common people did not travel, except for visiting the local markets. Goods were transported on foot or by the slow speed of carts:

> *Technically European agriculture was still, with the exception of a few advanced regions, both traditional and astonishingly inefficient. Its products were still mainly the traditional ones: rye, wheat, barley, oats, and in Eastern Europe buckwheat, the basic food of the people, beef cattle, sheep, goats and their dairy products, pigs, and fowl, a certain amount of fruit and vegetables, wine, and a certain number of industrial raw materials such as wool, hemp for cordage, barley for beer, etc. The food of Europe was still regional. The products of other climates were still rarities, verging on luxury, except perhaps for sugar, the most important foodstuff imported from the tropics and the one whose sweetness has created more human bitterness than any other* (Hobsbawm, 2010, pp. 17-18).

---

[16] The following general texts about changes are based on Wikipedia and other Internet sources. The information available on http://www.britannica.com on the History of Europe (particularly the chapter: "Revolution and the growth of industrial society, 1789–1914") was used extensively. Primary contributors to this topic are Hermann Aubin, Jacques Barzun, Timothy C. Champion, Michael Frassetto, David Herlihy, Judith Eleanor Herrin, Richard J. Mayne, N. Geoffrey Parker, Edward Peters, John Hearsey McMillan Salmon, Marie-Louise Stig Sorensen, Peter N. Stearns, Geoffrey Russell Richard Treasure, and Donald Weinstein. As their individual contributions cannot be traced, it was impossible to honor their work by citing them individually.

Major economic change was spurred by Western Europe's tremendous *population growth* during the late eighteenth century, extending well into the nineteenth century itself. Between 1750 and 1800, the populations of major countries increased between 50 and 100 percent, chiefly as a result of the use of new food crops (such as the potato) and a temporary decline in epidemic disease (pests, cholera). The population of Britain rose from 8.7 million in 1800 to 16.7 million in 1851 and 41.6 million in 1901. Population growth of this magnitude compelled change.

Figure 8: The Wood Sawyers
Painting by Jean-Francois Millet (1852).
Source: Wikimedia Commons, Wikiart

A full-scale technological revolution in the countryside occurred only after the 1850s. Nevertheless, factory-made tools spread widely even before this time, as scythes replaced sickles for harvesting, allowing a substantial improvement in productivity. Larger estates, particularly in commercially minded Britain, began to introduce newer equipment, such as seed drills for planting. Crop rotation, involving the use of nitrogen-fixing plants, displaced the age-old practice of leaving some land fallow, while better seeds and livestock and, from the 1830s, chemical fertilizers improved yields as well. (The first British patent for a chemical fertilizer was issued in 1842.) Rising agricultural production and market specialization were central to the growth of cities and factories. But the horse was still preeminently the power unit.

Peasant and artisanal children found their paths to inheritance blocked by sheer numbers and thus had to seek new forms of paying labor. Families of businessmen and landlords also had to innovate to take care of unexpectedly large surviving broods. In England these pressures occurred in a society already attuned to market transactions, possessed of an active merchant class, and blessed with considerable capital and access to overseas markets, as a result of existing dominance in world trade. Heightened commercialization showed in a number of areas. Vigorous peasants increased their landholdings—often at the expense of their less fortunate neighbors, who swelled the growing ranks of the near propertyless.

The peasants, in turn, produced food for sale in growing urban markets. Domestic manufacturing soared, as hundreds of thousands of rural producers worked full- or part-time to make thread and cloth, nails and tools under the sponsorship of urban merchants. Craft work in the cities

Figure 9: Horse-driven machinery (batteusse)

Source: Wikipedia Commons, Dictionnaire d'arts industriels (1881)

began to shift toward production for distant markets, which encouraged artisan-owners to treat their journeymen less as fellow workers and more as wage laborers. Europe's social structure changed toward a basic division, both rural and urban, between owners and nonowners. Production expanded, leading—by the end of the eighteenth century—to a first wave of consumerism as rural wage earners began to purchase new kinds of commercially produced clothing, while urban middle-class families began to indulge in new tastes, such as uplifting books and educational toys for children.

Technological changes

The nineteenth century was the period where the technological changes that started with the discoveries around the steam engine showed their effects in society.[17] Technological change was revolutionizing the life of the working classes. The Industrial Revolution that overtook England in the early eighteenth century spread rapidly over Europe. The substitution of human and animal labor with machine labor constituted the most important social changes seen in history. What would later be called the First Industrial Revolution started in England.

> *By any reckoning this was probably the most important event in world history, at any rate since the invention of agriculture and cities. And it was initiated by Britain. That this was not fortuitous, is evident. If there was to be a race for pioneering the Industrial Revolution in the eighteenth century, there was really only one starter…Whatever the British advance was due to, it was not scientific and technological superiority. In the natural sciences, the French were almost certainly ahead of the British; an advantage which the French Revolution accentuated very sharply, at any rate in mathematics and physics, for it encouraged science in France while reaction suspected it in England* (Hobsbawm, 2010 Chapter 2).

---

[17] See: B.J.G. van der Kooij: *The Invention of the Steam Engine* (2015).

Figure 10: Iron Rolling Mill by Adolf von Menzel (1875)
Source: Collection Nationalgalerie, Berlin

Within the European states, industrialization was producing new social classes: the industrial bourgeoisie and the industrial proletariat. Governments and private entrepreneurs worked hard to imitate British technologies after 1820, by which time an intense industrial revolution was taking shape in many parts of Western Europe—particularly in coal-rich regions such as Belgium, northern France, and the Ruhr area of Germany. And coal was important as it was the primary source of energy, feeding the boilers of the steam engine that drove the industrialization.

Technological change soon spilled over from manufacturing into other areas. Increased production heightened demands on the transportation system to move raw materials and finished products. Massive road and canal-building programs were one response, but steam engines also were directly applied as a result of inventions in Britain, Germany, and the United States. Steam ships plied major waterways soon after 1800, and by the 1840s local shipping had spread to oceanic transport. In the 1820s England railroad systems, first developed to haul coal from mines, were developed for intercity transport of persons and goods; the first commercial line opened between Liverpool and Manchester in 1830. During the 1830s local rail networks fanned out in most western European countries, and national systems were planned in the following decade, to be completed by about 1870. In the area of communication, the invention of the telegraph and telephone allowed faster exchange of news and commercial information than ever before.

New organization of business and labor was intimately linked to the new technologies. Workers in the industrialized sectors labored in factories rather than in scattered shops or homes. Steam and waterpower required a concentration of labor close to the power source. Concentration of labor also allowed new discipline and specialization, which increased productivity.

Social changes

At the beginning of the nineteenth century, the majority of people lived in the countryside or in small provincial towns:

> *The provincial town still belonged essentially to the economy and society of the countryside. It lived by battening on the surrounding peasantry and (with relatively few exceptions) by very little else except taking in its own washing. Its professional and middle classes were the dealers in corn and cattle, the processors of farm products, the lawyers and notaries who handled the affairs of noble estates or the interminable litigations, which are part of landowning or landholding communities; the merchant-entrepreneurs who put out and collected for and from the rural spinners and weavers; the more respectable of the representatives of government, lord, or church. Its craftsmen and shopkeepers supplied the surrounding peasantry or the townsmen, who lived off the peasantry*

(Hobsbawm, 2010, p. 12).

But that changed as people moved from the countryside to the cities to find work, and urbanization took place. It was a vital result of the ever-growing commercialization and the new industrial technologies. In England factory centers such as Manchester grew from villages into cities of hundreds of thousands in a few short decades. The population of London exploded form 0.96 million in 1801 to 2.36 million in 1851 and 6.53 million in 1901. The German Ruhr area changed within decades from an agricultural region into an industrial region. By around 1820 hundreds of water-powered mills were producing textiles, lumber, shingles, and iron in automated processes there. And in even more workshops in the hills, highly skilled workers manufactured knives, tools, weapons, and harnesses, using water, coal, and charcoal. By 1850 there were almost three hundred coal mines in operation in the Ruhr area, in and around the central cities of Duisburg, Essen, Bochum, and Dortmund.

Figure 11: Spinner in Vivian Cotton Mills, Cherryville, N.C. (1908)

Source: http://historyinphotos. blogspot.fr/2012/07/lewis-hine-mill-workers-ctd.html

The percentage of the total population located in cities expanded steadily, and big cities tended to displace more scattered centers in Western Europe's urban map. Rapid city growth produced new hardships, for housing stock, and sanitary facilities could not keep pace, though innovation responded, if slowly. Gas lighting improved street conditions in the better neighborhoods from the 1830s onward, and sanitary reformers pressed for underground sewage systems at about this time. For those who were better-off, rapid suburban growth allowed some escape from the worst urban miseries. Rural life changed less dramatically.

But it was the working conditions that changed most considerably. Wage laborers' autonomy of work declined; more people worked under the daily direction of others. Early textile and metallurgical factories set shop rules, which urged workers to be on time, to stay at their machines rather than wandering around, and to avoid idle singing or chatter (difficult in any event given the noise of the equipment). These rules were increasingly enforced by foremen, who mediated between owners and ordinary laborers. Work speeded up. Machines set the pace, and workers were supposed to keep up. The nature of work shifted in the propertied classes as well. Middle-class people, not only factory owners but also merchants and professionals, began to trumpet a new work ethic. According to this ethic, work was the basic human good. He who worked was meritorious and should prosper; he who suffered did so because he did not work.

The growth of cities and industry had a vital impact on family life. The family declined as a production unit as work moved away from home settings. This was true not only for workers but also for middle-class people. Many businessmen setting up a new store or factory in the 1820s initially assumed that their wives would assist them, in the time-honored fashion in which all family members were expected to pitch in. After the first generation, however, this impulse faded, in part because fashionable homes were located at some distance from commercial sections and needed separate attention.

Figure 12: Frame-breakers, or Luddites, smashing a loom (1812)
Source: Wikimedia Commons

In general most urban groups tended to respond to the separation of home and work by redefining gender roles, so that married men became the family breadwinners (aided, in the working class, by older children), and women were the domestic specialists tending the numerous offspring.

Class divisions manifested in protest movements. Middle-class people joined political protests hoping to win new rights against aristocratic monopoly. Workers increasingly organized on their own, despite the fact that new laws banned craft organizations and outlawed unions and strikes. Some workers attacked the reliance on machinery in the name of older, more humane traditions of work. Luddite protests of this sort began in Britain during the decade 1810–20.[18] (See Figure 12.) More numerous were groups of craft workers, and some factory hands, who formed incipient trade unions to demand better conditions as well as to provide mutual aid in cases of sickness or other setbacks. Social protest was largely intermittent because many workers were too poor or too disoriented to mount a larger effort, but these protests clearly signaled important tensions in the new economic order.

> *Its most serious consequences were social: the transition to the new economy created misery and discontent, the materials of social revolution. And indeed, social revolution in the form of spontaneous risings of the urban and industrial poor did break out, and made the revolutions of 1848 on the continent, the vast Chartist movement in Britain. Nor was discontent confined to the laboring poor. Small and inadaptable businessmen, petty-bourgeois, special sections of the economy, were also the victims of the Industrial Revolution and of its ramifications. Simple-minded laborers reacted to the new system by smashing the machines, which they thought responsible for their troubles; but a surprisingly large body of local businessmen and farmers sympathized profoundly with these Luddite activities of their laborers, because they too saw themselves as victims of a diabolical minority of selfish innovators. The exploitation of labor, which kept its incomes at subsistence level, thus enabling the rich to accumulate the profits which financed industrialization (and their own ample comforts), antagonized the proletarian…*
>
> *It was all very well for the rich, who could raise all the credit they needed, to clamp rigid deflation and monetary orthodoxy on the economy after the Napoleonic Wars: it was the little man who suffered, and who, in all countries and at all times in the nineteenth century demanded easy credit and financial unorthodoxy. Labor and the disgruntled petty-bourgeois on the verge of toppling over into the unpropertied abyss, therefore shared common discontents* (Hobsbawm, 2010, pp. 38-39).

---

[18] The Luddites were nineteenth-century English textile artisans who protested against newly developed laborsaving machinery from 1811 to 1817. They advocated sabotage (by throwing their wooden clogs—*sabot* in French—into the gears of textile machines). See: (Hobsbawm, 1952).

Political changes [19]

Looking at the beginning of the nineteenth century, the French Revolution had had its effect, and "power came to the people." This was not totally true though as the old powers (aristocracy, clergy) did not surrender their interests that easily. Europe was under French influence, and the French political concepts (*"Liberté, Egalité, Fraternité"*) and French laws were introduced.

After the collapse of the First French Empire, the victorious powers convened at the Congress of Vienna (September 1814 to June 1815) to try to put Europe back together. These victors were of the opinion that there should be no thought of literally restoring the world that had existed before 1789 (Figure 13). Regional German and Italian states were confirmed as a buffer to any future French expansion. Prussia gained new territories in western Germany. Russia took over most of Poland. Britain acquired some former French, Spanish, and Dutch colonies (including South Africa). The Bourbon dynasty was restored to the French throne in the person of Louis XVIII, but revolutionary laws were not repealed, and a parliament—though based on very narrow suffrage—proclaimed a constitutional monarchy.

Next to these political changes on the level of nations, there were also political changes within those nations themselves. Here the forces that changed society from medieval feudalism into monarchial and papal dominance, played their own and different roles.

As these old players tried to conserve and protect their positions, "conservatism" dominated the European political agenda through the mid-1820s. Major governments, even in Britain, used police agents to ferret out agitators. The prestige of the Roman Catholic Church soared in France and elsewhere. Liberal agitation began to revive in Britain, France, and the Low Countries by the mid-1820s. Liberals wanted stronger parliaments and wider protection of individual rights. They also sought a vote for the propertied classes. They wanted commercial legislation that would favor business growth, which in Britain meant attacking Corn Law tariffs that protected landlord interests and kept food prices artificially high. Belgian liberals also had a nationalist grievance, for the Treaty of Vienna (1815) had

---

[19] Politics are the mechanism of power in social interactions between people in social groups, societies, and nations. Politics are related to ones beliefs about humanity (liberalism, socialism, conservatism), the role and position of the individual in the group (company, state) and the promoting and implementation of these believes. As politics in democracies mean the division of power, it also includes negotiating. The result being the influence the state realizes on the society by laws and policies. It also can include the exercise of force on individual level (police) or state level (military) in the case of political conflicts (uprisings, wars).

placed their country under Dutch rule—a fact that was corrected when the Belgian Revolution (1830) resulted in the creation of Belgium.

The nineteenth century was a time in which liberalism, based on liberty and equality, spread all over Europe creating liberal governments and liberal democracies. The rise of the middle class, who wanted their rights, changed the power balance between aristocracy and the working class.

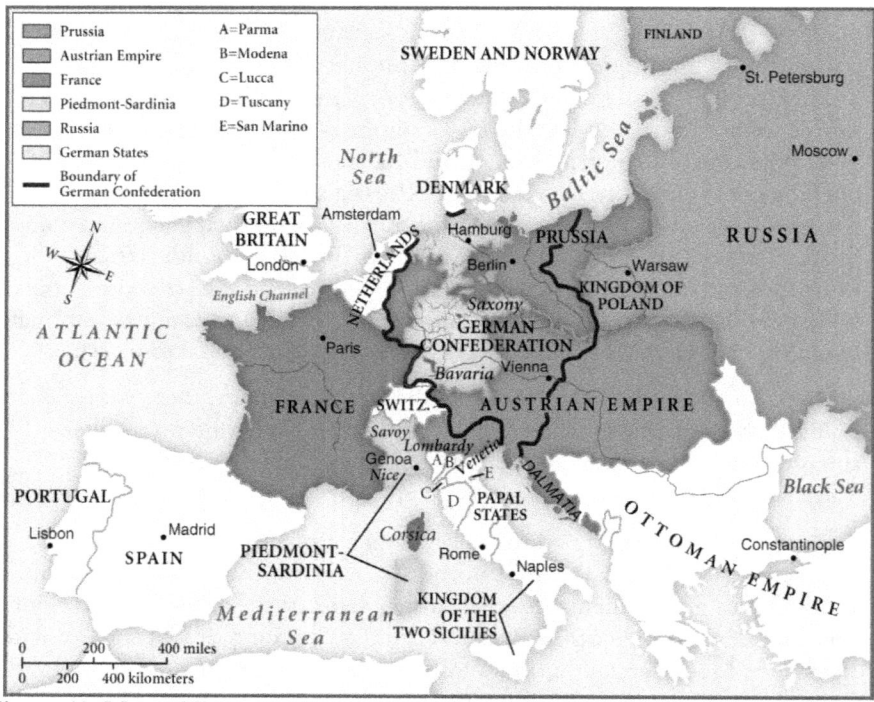

Figure 13: Map of Europe, after the Congress of Vienna (1815)
Source: Wikimedia Commons

Climate and the affairs of man[20]

The abovementioned changes in European societies—technical, economic, social, and political—were related to, and certainly the result of, people's individual behavior within the context of their societies. However, there was another dominant factor that influenced this human behavior. That factor was nature itself with its climate conditions that resulted in bad harvests and periods of famine. For example the Great Famine in Ireland

---

[20] See also: Winkless, Nels III and Browning, Iben, "Climate and the Affairs of Men," *Harper's Magazine Press*, 1975.

and Scotland was caused by a potato disease (the fungus "phytophthora infestans"), that resulted in mass starvation, disease, and emigration between 1845 and 1852. More than a million people died and a million people emigrated. The European potato failure also affected the lowlands—Belgium, especially Flanders, the Netherlands, Prussia, and France. It was also a period of poor wheat and rye harvests throughout much of Europe. It resulted in the subsistence crisis of 1845–1850.

> *The harvest of 1846 was a different story. In Ireland and the Scottish Highlands the potato yields were barely enough or insufficient to provide for the next year's seed. On the other hand, in 1846 potato yields improved in Belgium and The Netherlands. However, in these countries and also in Prussia about half the potato harvest was lost. In much of northwestern Europe the problems caused by the potato were exacerbated by the loss in 1846 of almost half of the rye harvest, while the wheat harvest was considerably below normal. This was disastrous, with bread from rye or wheat being even more important than potatoes in continental European diets. It bears emphasis that the failure of the Dutch, Belgium, and Prussian rye harvest of 40 to 50 percent in 1846 was extreme by nineteenth century standards* (Vanhaute et al., 2007, p. 11).

This all resulted in considerable economic disturbances. As food was scarce, prices went up, and the poor could less afford their basic nutrition. People rioted and criminality rose.

> *A real wave of market disturbances surged over Europe in 1846–1847, with a top in spring (April, May, June) 1847, when grain prices peaked. It is striking that regions with market-oriented agriculture and a substantial number of wage laborers were by far most affected by market disturbances. In France riots were heaviest in cities and in grain exporting regions…the pattern in Spain is very similar: a huge wave of short time market riots (mostly lasting one or two days) in the first half of 1847, all instigated by an (expected) rise of grain and bread prices and (presumed) maneuvers of speculation and export…In other regions, such as South Germany, Flanders, and The Netherlands, riots were almost exclusively urban events, mostly directed against the symbols of (perceived) speculation, such as millers, bakers, and traders…Rising unrest was reflected in higher numbers of registered criminality (such as theft, cattle and sheep rustling, burglaries, and robberies). In Flanders small food riots were confined to the cities. On the other hand, petty criminality (notably mendicancy and vagrancy, petty theft, pillage, stealing crops) rose by 50 percent in the crisis years 1846–1847* (Vanhaute et al., 2007, pp. 17-18).

As people were not able to meet their nutritional needs, their physical situation deteriorated, their health weakened, and they became more vulnerable to diseases—diseases such as the cholera pandemics that reached Europe as a result of the increasing trade contacts in the 1816–1851 time frame and claimed hundreds of thousands of lives.

> *There can be little doubt that cholera epidemics tended to occur at moments of crisis in European history. The first great epidemic came as the reverberations of the revolutions of 1830 were still echoing across the Continent, reaching Britain during the profound political crisis over the Great Reform Bill of 1832, a year which, some historians have argued, saw the only real possibility of a political revolution in modern British history. Cholera next swept across Europe in the revolutionary year of 1848—probably the greatest and most devastating of all the epidemic years, at least in terms of the sheer number of people affected...A second cause of mass movements of human beings was indeed famine and deprivation. In 1847–8 in particular, hunger and destitution drove vast numbers of people in central Europe to flee from the countryside to the towns, and from one town to another, in search of poor relief...* (R. J. W. Evans, 1988, pp. 131, 133).

The shortfalls in agricultural output led to a decline in manufacturing, increased imports, and rising food prices. Obviously the scarcity of goods did give way to speculation, thus increasing the prices again. People spent most of their income on food and restrained from buying other products of a less vital nature. All this resulted in periods of economic depression and in commercial crisis with its many bankruptcies of many mercantile houses—bankruptcies that, in turn, had their effects on the monetary system—for example the British banking crisis of 1847 (D. M. Evans, 1849).

> *Meanwhile wheat prices were soaring to levels unknown since 1817, and in May [1846] they touched 112s. per quarter. But it was soon clear that the dealers had grossly underestimated the elasticity of supply, while to add to their discomfiture harvest prospects for 1847 were good. After the peak, prices came tumbling down, and in August the failure of corn dealers in both London and Liverpool started. These in their turn involved other houses, who had extended them credit, and like a house of cards, the overstrained credit structure collapsed. The pressure towards liquidity became more intense; bills even of a first-class nature became increasingly difficult to cash. As the crisis developed, the Bank manfully attempted to fulfill the role of lender in the last resort, and week by week the reserve in the Banking Department dwindled. The real panic came when it was obvious that this reserve was becoming exhausted, and that under the provisions of the Bank Charter Act the Bank would have to refuse further advances and discounts. Everyone,*

including the soundest houses, scrambled for liquidity and assets other than bank notes were almost unmarketable; the heaviest rates gave no incentive to lenders (Ward-Perkins, 1950, p. 78).

The relation between these natural disasters—with their diseases, the resulting agricultural decline, and, consequently, the following economic and financial distress—and the social unrest and political changes that occurred during and after the European Revolutions of 1948 is evident.

In sum, it would seem that the deterioration of financial conditions in the wake of the agrarian crisis of 1845–1847 had a sizable lagged impact on firm failures and investment behavior, which transmitted the crisis across sectors and into the critical year 1848…Our findings thus support the idea that the character of the regime largely determined the form of political upheaval in 1848, but that it was the economic crisis that set the wheels of revolution in motion (Berger & Spoerer, 2001, pp. 306, 318).

The Second Industrial Revolution

The sum of all these changes, from technical to social, economic and political, created the basis for the Second Industrial Revolution. In the first

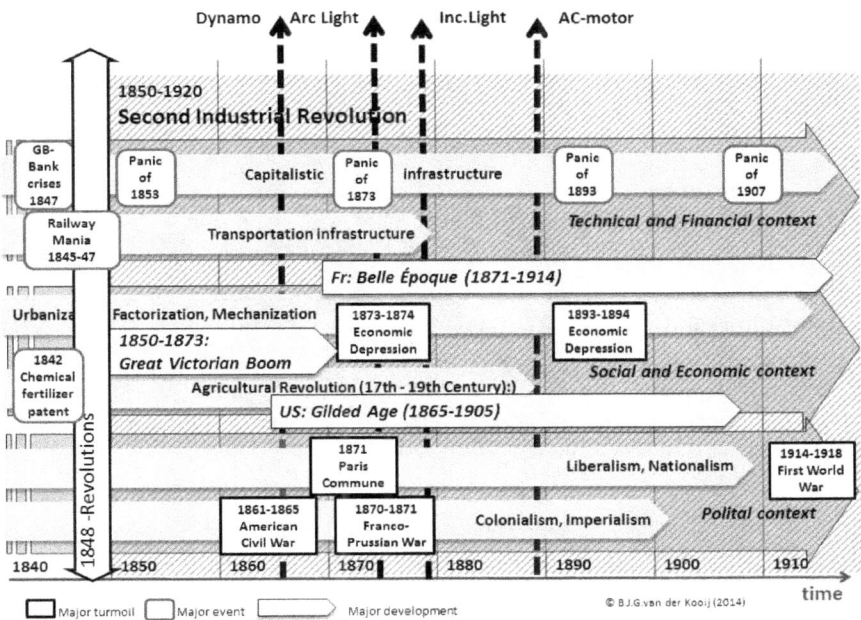

Figure 14: The context that would result in the Second Industrial Revolution
Source: Figure created by author

half of the nineteenth century the First Industrial Revolution, initiated by the technical changes induced by the steam engine[21], laid the foundations for a massive change that would occur in the second half of the nineteenth century. It would, in the 1840s-1850s evolve in the Second Industrial Revolution.

It is within this political, social, economic, technical and financial context that the new electric technologies developed, both in Europe and in the United States (Figure 14).

Science discovers and applies electricity

Curious people in history observed and discovered, often more or less by accident, phenomena they could not explain, but—in hindsight—they were encountering basic elements of electricity. Many of those curious people, also called the "gentlemen of science," were wealthy and independent—a position that enabled them to spend their time on matters that sparked their curiosity and to join institutions such as the Academie Royale des Science in Paris, the Royal Society of London, the Academy Royale of Berlin, and the Academy of Sciences of St. Petersburg.

The electrophysicists

Take the English physicist William Gilbert (1544–1603) who discovered the phenomenon of the "amber[22] effect"—the electric charge that appears when substances such as amber, sulfur, and glass are rubbed (Schiffer, 2006, pp. 14-17) (Heilbron, 1979, pp. 167-169). It was Otto von Guericke (1602–1686) who constructed an electrical device when he made his glass

Figure 15: Hauksbee's glass-globe style electrical machine

Source: Wikimedia Commons, drawing by French clergyman and scientist Jean-Antoine Nollet, from his 1767 book Leçons de Physique.

---

[21] See: B.J.G. van der Kooij: *The Invention of the Steam Engine* (2015)
[22] The Greek word for amber is electrum.

globe of 15 cm in diameter filled with sulfur. After heating the glass and melting the sulfur, he ended up with a sulfur globe that, when rubbed by hand, could be used to create frictional electricity that attracted light objects. He had created an electrostatic generator (Schiffer, 2006, pp. 17-19). And Francis Hauksbee (1666–1713), Isaac Newton's lab assistant and chief experimentalist at the Royal Society of London, made a machine that created electric charge. He presented it at the Royal Society on December 5, 1703 (Schiffer, 2006, pp. 23-26).

> *In France electricity also had caught the attention. The Frenchman Charles Francois de Cisternay du Fay (1698–1739), also known as Dufay, member of the Paris Academy of Science, independently wealthy, equally at home among academicians, ministers, and high society, became the successor of Hauksbee's work (Heilbron, 1979, pp. 250-260). Dufay, after experiments with the "universal discharger," an electrical machine with a hand-rubbed glass globe, concluded in 1733 that electricity comes in two varieties: one produced by glass, the vitreous charge, and one produced by resin, the resinous charge (later known as positive and negative charge). And he concluded that objects possessing the same charge repel each other, and those having an opposite charge, attract each other. In 1734 he published his two-fluid concept in the annals of the Royal Society of London: "Two Kinds of Electrical Fluid: Vitreous and Resinous"* (du Fay, 1734).

This two-fluid concept was later, in 1769, adapted by Benjamin Franklin as two states of the same single electricity (plus and minus[23]) of an electrical fluid. Franklin's letters about the subject were also published by the Royal Society (Schiffer, 2006, pp. 30-31,48).

Other experimenters who created electrical machines were the Frenchman Jaen-Antione Nolett (1700–1770), working for DuFay and also a believer in the two-fluid theory of electricity; the Dutchman Pieter van Musschenbroeck (1692–1761), professor at the University of Leiden, who created—together with his brother who was an instrument maker—several electrical devices; and

Figure 16: Leyden jars (Eighteenth century)
Source: www.sparkmuseum.com

---

[23] Positive was considered to be a surplus (+), and negative was considered a deficiency of electrical fire. Today we consider "negative" as the surplus of electrons, the carriers of negative charge. Even so, "positive" is a deficiency of electrons.

in Germany Gerg Bose (1710–1761) of the Wittenburg University, who improved on Hauksbee's electrical machine (Figure 15) (Schiffer, 2006, p. 39). All in all, one can conclude that with these men's discoveries, the first electricity-*generating* device, the electrostatic machine, was born.

> *An interesting discovery—by accident—was that of the "Leyden Jar," a device that "stores" static electricity between two electrodes on the inside and outside of a glass jar. It was the original form of a capacitor. In many laboratories, glasslike containers were available. It were Ewals Jurgen von Kleist (1700–1748) and van Pieter van Musschenbroeck who both noticed that the jars could be electrically charged and gave them a shock (Heilbron, 1979, pp. 307-316). It was Benjamin Franklin who started to use the Leyden Jars in parallel and named it a "battery"* (Schiffer, 2006, pp. 45-47).

> *It was a range of people who more or less observed the same phenomenon and tried to explain it. These "electrophysicists" as they were called, had discovered the "electric charge" of static electricity, a knowledge that was rapidly transferred to other scientific communities: atmospheric electricians and electrochemists, but also by communities outside science, including the property protectors, electrical collectors, electrotherapists, and electrical demonstrators*
> (Schiffer, 2002, p. 1158).

So in the eighteenth century, science had progressed and got a first grasp on the new phenomenon "electricity." It was already recognized as being important as formulated by Samual Klingenstierna, professor of physics at the University in Uppsala, in 1755:

> *Forty years ago, when one knew nothing about electricity but its simplest effects, when it was regarded as an unimportant property of a few substances, who would have believed that it could have any connection with one of the greatest and most considerable phenomenae in Nature, thunder and lightning?*
> (Heilbron, 1979, p. 6).

Joseph Priestly (1733–1804)

One of the early scientists who participated in the emerging fields of chemistry and electricity was Joseph Priestly (1733–1804). He was intended for ministry and learned Greek, Latin, and Hebrew at an early age. A serious illness in 1749 left him with a stutter, and he gave up any thoughts of entering ministry at that time. He next studied French, Italian, and German, and then Chaldean, Syrian, and Arabic. In 1752 he matriculated at

Daventry, a dissenting academy, and became a Rational Dissenter.[24]

In 1761, at the age of twenty-six, Priestley moved to Warrington and assumed the post of tutor of modern languages and rhetoric. The intellectually stimulating atmosphere of Warrington, often called the "Athens of the north" during the eighteenth century, encouraged Priestley's growing interest in natural philosophy. Despite Priestley's busy teaching schedule, he decided to write a seven-hundred-page history of electricity: *The history and present state of electricity: with original experiments* (Priestley, 1769b). This resulted in his membership of the Royal Society on June 12, 1766. The study of "electrostatics" had become one of the most popular of the Newtonian sciences during the first half of the eighteenth century. And Priestley's text became the standard history of electricity for over a century. Alessandro Volta (who later invented the battery), William Herschel (who discovered infrared radiation), and Henry Cavendish (who discovered hydrogen) all relied upon it. Between 1767 and 1770, Priestley presented five papers to the Royal Society and the popular version of his history of electricity: *A Familiar Introduction to the Study of Electricity* (Priestley, 1769a).

In 1767 Priestley became a minister in Leeds. There he wrote *Institutes of Natural and Revealed Religion* (Priestley, 1782). He combined scientific thinking with religion, and because of this he became engaged in numerous political and religious pamphlet wars (Priestley, 1771). When he moved to Leeds, Priestley continued his electrical and chemical experiments. In 1772 William Petty Fitzmaurice, First Marquess of Lansdowne, Second Earl of Shelburne, and prominent politician (in short Shelburne), asked Priestley to direct the education of his children and to act as his general assistant (his "literary companion"). Priestley would receive £250 per annum,[25] the use of a house in Calne (two miles from Bowood, the residency of Shelburne), and an annuity of £150 for life when their ways would part (Rivers & Wykes, 2008).

> *[Shelburne] encouraged me in the prosecution of my scientific enquiries and allowed me 140 per annum for expenses of that kind, and was pleased to see me make experiments to entertain his guests, and especially for foreigners* (Griffith, 1983, p. 7).

---

[24] The English Dissenters were Christians who separated from the Church of England. Dissenters opposed state interference in religious matters and founded their own churches, educational establishments, and communities. They could not hold political office, serve in the armed forces, or attend Oxford and Cambridge unless they subscribed to the Thirty-nine Articles of the Church of England. Abhorring dogma and religious mysticism, Rational Dissenters emphasized the rational analysis of the natural world and the Bible.

[25] Calculated on the historic standard of living value, that income would be equivalent to £24.930 in 2010. Source: http://www.measuringworth.com/ukcompare/relativevalue.php.

In 1772 he published his *The History and Present State of Discoveries Relating to Vision, Light and Colours*. In 1773 Priestley moved to Calne. Priestley's years in Calne were the only ones in his life dominated by scientific investigations; they were also the most scientifically fruitful. In August 1774 he isolated an "air" that appeared to be completely new, but he did not have an opportunity to pursue the matter because he was about to tour Europe with Shelburne. From August 24, 1774, to November 2, 1774, Priestley accompanied Shelburne on a trip to Flanders, Holland, Germany, and France. While in Paris he met Lavoisier and other chemists and managed to replicate his experiment for others, including French chemist Antoine Lavoisier. After returning to Britain in January 1775, he continued his experiments and discovered "vitriolic acid air" (sulfur dioxide, SO_2). Later he discovered oxygen (O_2), a discovery that he had to share with the Swedish scientist Carl Scheele (Priestley & DFRS, 1775). This was during the time of the American War of Independence (1775–1783) in which the United States became independent of British rulers.

> *In 1779 Priestley heard that Shelburne was considering transferring him to one of his Irish estates. With a sure sense of history, he took this as a hint that their association was about to end—it is possible that his theological views had become too radical for his patron....Shelburne paid him an annual pension of £150*[26] *to the end of his life when he again became dependent on other patrons*

(Griffith, 1983, p. 8).

After this rupture with Lord Shelburne, a politically active man who became Prime Minister in 1782, the Priestleys moved to Birmingham. Birmingham had become quite industrialized, with many small independent domestic workshops, many poor laborers, and a growing bourgeois class of merchants with an elite of a few industrial families (Rose, 1960, p. 70).

In the 1780s Priestley spent a happy decade surrounded by old friends. He became a member of the Lunar Society, a group of manufacturers, inventors, and natural philosophers who assembled monthly to discuss their work. There he met with Matthew Boulton and James Watt. Priestley published several more scientific papers in Birmingham, the majority attempting to refute Lavoisier's new concepts of chemistry. In 1783 he published *Experiments relating to Phlogiston, and the seeming Conversion of Water into Air* (Priestley, 1786). He wrote more publications, became engaged in pamphlet wars in which he was declared an atheist, and supported the French Revolution. Dissenters such as Priestley who supported the French Revolution came under increasing suspicion as skepticism regarding the

---

[26] Calculated on the historic standard of living value, that income would be equivalent to £15.530 in 2010. Source: http://www.measuringworth.com/ukcompare/ relativevalue.php.

revolution grew (Figure 17). The animus that had been building against Dissenters, supporters of the American rebellion against England, and the French Revolution exploded in July 1791. The resulting violence became known as the "Church and King" riots.

Figure 17: The Treacherous Rebel cartoon showing Priestley
Source: Wikimedia Commons/Timmins Collection

Within the pattern of Birmingham politics in the eighteenth century the questions of the Test and Corporation Acts[27] assumed a peculiar importance. The agitation for repeal that began in 1787 disturbed the established equilibrium; it led the dissenters to form separate, sectarian groupings and these in turn provoked Anglican hostility…The Church did exert itself, however, and a meeting was held at Warwick on 2 February 1790, attended by the "noblemen, gentlemen, and clergy" of the county, to concert measures of opposition to repeal. A local committee was also formed to conduct opposition to the dissenters in Birmingham. The pamphlet war continued throughout the year, and in December 1790 relations became so strained that the magistrates obtained the sanction of the War Office for the dispatch of Dragoons from Derby and Leicester, in case rioting seemed imminent (Rose, 1960, pp. 71, 72).

Priestley and several other Dissenters had arranged to have a celebratory dinner at the Royal Hotel on July 14, 1791, the anniversary of the storming of the Bastille—quite a provocative action in a country where many disapproved of the French Revolution and feared that it might spread to Britain. Amid fears of violence, Priestley's friends convinced him not to attend. Rioters gathered outside the hotel during the banquet and attacked the attendees as they left. The rioters moved on to the New Meeting and Old Meeting churches—and burned both to the ground.

From the Old Meeting the rioters went to Dr. Priestley's house at Fair Hill, Sparkbrook, which they ransacked and then burned, destroying an immensely valuable collection of manuscript material and apparatus. Priestley managed his personal escape with little time to spare (Rose, 1960, p. 73).

---

[27] The Test Acts were a series of English penal laws that served as a religious test for public office and imposed various civil disabilities on Roman Catholics and Nonconformists.

Although Priestley and his wife fled from their home, Fairhill at Sparkbrook, their son William and others stayed behind to protect their property; however, the mob overcame them and torched Priestley's house, destroying his valuable laboratory and all of the family's belongings. In his later publication, *An Appeal to the Public on the Subject of the Late Riots in Birmingham* (Figure 18), he wrote about the event:

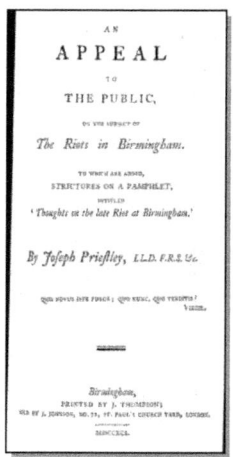

Figure 18: Title page from *An Appeal to the Public on the Subject of the Riots in Birmingham* (1791)
Source: Wikimedia Commons

> *It being remarkably calm, and clear moon-light, we could see to a considerable distance, and being upon a rising ground, we distinctly heard all that passed at the house, every shout of the mob, and almost every stroke of the instruments they had provided for breaking the doors and the furniture. For they could not get any fire, though one of them was heard to offer two guineas for a lighted candle; my son, whom we left behind us, having taken the precaution to put out all the fires in the house, and others of my friends got all the neighbours to do the same. I afterwards heard that much pains was taken, but without effect, to get fire from my large electrical machine, which stood in the library* (Priestley & Johnson, 1792, p. 30).

The riot continued the next day, and the town prison was attacked and prisoners released. In the afternoon and evening of the fifteenth, several more houses were attacked, and all business was suspended in the town.

> *The centre of the town was still in a lawless state, however, and "in the afternoon and evening small parties of three or five levied contributions of meat, liquor, and money…with the same indifference that they would levy parish taxes." Business was "at a stand" all day "and the shops mostly close shut up."…The rioters appear to have considered three classes of person as legitimate targets. The reformers who attended the Bastille dinner were the first to suffer; dissenters of various denominations were then attacked, and finally, members of the Lunar Society, in which Boulton, Watt, Priestley, Keir, and Withering were prominent* (Rose, 1960, pp. 75, 76).

Priestley spent several days hiding with friends until he was able to travel safely to London. From there he went to live in France in June 1792. There he was declared, together with George Washington and others, in absentia a member of the National Convention in France (Griffith, 1983, p. 10).

> *The net result of the riots was to create a catastrophic new division in the political life of Birmingham. James Watt commented in November 1791 that "the town is divided into two parties who hate one another mortally, that the professed aristocrats are democrats in practice, that is, encouragers of the Mob; and that the democrats are those who have always contended for a police and good government in the town, therefore are in fact aristocratic, at least would have no objections to an aristocracy of which they themselves were member…" It may be that the Priestley riots ought to be regarded in retrospect as an episode in which the "country gentlemen" called out the urban mob to draw the dissenting teeth of the aggressive and successful Birmingham bourgeoisie* (Rose, 1960, pp. 83, 84).

As the penalties became harsher for those who spoke out against the government, and despite his election to the French National Convention by three separate departments in 1792, Priestley decided to move with his family to the United States.

> *In late August 1793, his sons Joseph and Henry sailed for America. Priestley waited to receive part of the £3098[28] that he was finally to be paid for compensation for damages sustained during the Birmingham Riots and then decided to follow them with his wife. He gave his farewell sermon on 30 March 1794 at the Gravel Pit and spent his last days in England with Lindsey, his last Sunday being spent at the Essex Street Chapel. He sailed on the Sansom from Gravesend on 8 April 1794, finally reaching America on Wednesday 4 June* (Griffith, 1983, p. 10).

Five weeks after Priestley left, William Pitt's administration (Pitt being the successor of Sheldon) began arresting radicals for seditious libel, resulting in the famous 1794 Treason Trials. By 1801 Priestley had become so ill that he could no longer write or experiment. He died on the morning of February 6, 1804, in Northumberland, Pennsylvania.

This rather extensive description of Priestley's life illustrates the context in which scientists were living and working. Science was not isolated from religion nor from earthly matters such as social and political change and its related turmoil.

---

[28] In 2010 that would be more than $283,400 (based on the historic standard of living calculation). Source: http://www.measuringworth.com/uscompare/relativevalue.php.

Electricity as phenomenon: the nature of lightning

Electricity is not something created by people; it is part of our natural environment. Mankind has been confronted with it every time a thunderstorm creates "electric lightning." In older times people believed lightning was caused by Thor (Donar or Wodan) who was throwing his hammer. Nowadays we know that it is an electric charge, accumulated by moving particles in the clouds, which is discharged by a current of electrons traveling to the earth.[29] This discharge and the effects of this massive flow of electrons (the electric current) is quite dramatic (noise, light), and it can have destructive consequences (death, destruction, fires). It is called the electromagnetic force, one of the four fundamental forces of nature (the others being gravitation, strong nuclear, and weak nuclear). This phenomenon of lightning has sparked many people's curiosity over time as they wondered what the mechanism was behind the "power of lightning."

Figure 19: Lightning, the discharge of static electricity
Source: Wikimedia Commons

In the eighteenth century, when the "gentlemen of science" focused their attention on the basic mechanisms of our natural environment (such as the "power of fire" as described elsewhere[30]), electricity became an area of interest. It was the work of these eighteenth century scientists that created the foundations for later developments. The experiments of people such as the Italians Luigi Galvani (1737–1798) and Alessandro Volta (1745–1827), the Frenchmen Andre-Marie Ampère (1775–1836) and Charles-Augustin de Coulomb (1736–1806), the Brit Joseph Priestley (1733–1804), and the American Benjamin Franklin (1706–1790) gave us insights into the mechanism of electricity—for example the form called "static electricity."

Nature of lightning: static electricity

Benjamin Franklin—a so-called "atmospheric electrician"—was highly interested in the phenomenon of atmospheric electricity as it appeared in lightnings in thunderstorms. Together with friends he was interested in the meaning of these manifestations of "the electric fire." What we call electricity nowadays was considered to be a fluid in those days. Some

---

[29] Electricity concerns moving electrons—parts of atoms with a negative electrical charge. Atoms are parts of molecules that create our universe.
[30] See: B. J. G Van der Kooij. *The Invention of the Steam Engine* (2015).

considered it as two fluids (as did the Frenchman Charles du Fay, 1698–1739), but Franklin saw it differently:

> The new one-fluid conception of electricity gave Franklin an insight into many complex electric phenomena, including the condensing property of the Leyden jar,[31] and was of course an anticipation of modern ideas on the electrical structure of matter, in which electrons, detached from atoms, comprise the "subtile fluid"
>
> (B.F.J. Schonland, 1952, pp. 376-377).

He developed an electrical machine with which he could charge a conducting body. And his experiments resulted in the lightning rod, which protected buildings from lightning bolts.

Franklin became famous for bringing lightning down to earth with his Philadelphia experiments (Figure 20). In his experiment in 1750, he proved the existence of electricity by flying a kite in a thunderstorm. The kite twine conducted the "electric fire" along the twine to a key at the bottom. Franklin wrote in a letter to his friend Peter Collins of London:

Figure 20: Benjamin Franklin's kite experiment
Source: www.americaslibrary.gov

> When rain has wet the kite twine so that it can conduct the electric fire freely, you will find it streams out plentifully from the key at the approach of your knuckle, and with this key a phial, or Leiden jar, may be charged: and from electric fire thus obtained spirits may be kindled, and all other electric experiments [may be] performed which are usually done by the help of a rubber glass globe or tube; and therefore the sameness of the electrical matter with that of lightening completely demonstrated (Franklin, 1751, p. 566).

Other scientists repeated his experiments, such as D'Alibard in France in 1752 and Georg Wilhelm Richmann in 1753 in St. Petersburg. A 'ball of lightning' electrocuted the latter, who experimented by conducting lightning

---

[31] The Leyden jar was a device that could hold an electrical charge; we would call it an electric condenser nowadays. It was discovered in 1745 by the German E. G. von Kleist (1700–1748) and the Dutchman P. van Musschenbroek (1692–1761).

bolts into an isolated rod. It was the Brit Joseph Priestley who described his discovery a few years later as "the greatest, perhaps, that has been made in the whole compass of philosophy since the time of Sir Isaac Newton" (B. F. J. Schonland, 1952, p. 380).

So Franklin and others established the existence of static electricity: the "electrical fire." Since static electricity can also be created by friction or rubbing, it was also termed "frictional electricity." For a long time, the theories of the dual-fluid concept versus the single-fluid concept caused intense debates among the scholars (Heilbron, 1979, pp. 431-448). But next to the electric fire there was to be discovered another type of electricity: the animal electricity.

Nature of lightning: animal electricity

During this same time period, there had also been scientists looking for another form of electricity called "animal electricity." Take the famous example of Luigi Galvani, professor of anatomy at the University of Bologna, who in 1786 discovered that a frog's legs would exhibit violent muscular contractions when its exposed nerves were touched with one metal and its muscles were touched with another metal, while the two metals were connected.[32] Galvani explained this occurrence as the discharge of the "nerveo-electrical fluid" previously accumulated in the muscle (Heilbron, 1979, p. 491).

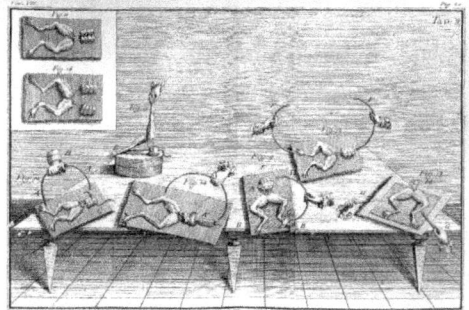

Figure 21: Galvani's experiment with frog legs (1791)

Source: Wikipedia Commons, Commnetarius

His discovery was the result of a range of experiments with electricity. Aware of static electricity from lightning, Galvani started his experiments with observing the contractions due to atmospheric electricity during a weather storm.

> He connected the frog nerve to a long metallic wire pointing toward the sky, in the highest place of his house and "...in correspondence of four thunders, contractions not small occurred in all muscles of the limbs, and, as a consequence, not small

---

[32] Now we know that the effect was due to a (very small) electric current generated by a chemical reaction and acting with contractile effect on the muscles of the frog's legs.

hops and movements of the limbs. These occurred just at the moment of the lightnings; they occurred well before the thunders when they were produced as a consequence of these ones" (Piccolino, 1998, p. 385).

Next he repeated the experiment during serene weather in order to investigate if the natural electricity present in the atmosphere of a calm day could succeed in evoking contractions:

> *But nothing happened for a long time. Finally "tired of the vain waiting" he came near the railing and started manipulating the frogs. To his great surprise, the contractions appeared when he pushed and pressed, toward the iron bars of the railing, the metallic hooks inserted into the frogs spinal cord (Ibidem).*

Galvani repeated his experiments indoors (Figure 21) "and realized that, in order to get contractions, it sufficed to connect through a metallic conductor the nervous structures (crural nerves or spinal cord) and the leg muscles, therefore, creating a circuit 'similar to that which develops in a Leyden jar'…when the internal and external plates are connected…Galvani came to the conclusion that some form of intrinsic electricity was present in the animal, and that connective nerve and muscle together, by means of conductive materials, induced contractions by allowing for the flow of this internal electricity" (Piccolino, 1998, p. 386).

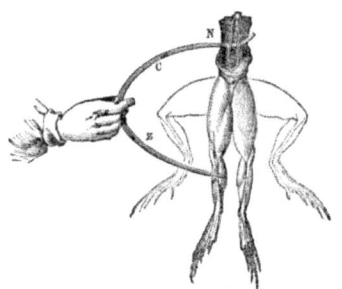

Figure 22: Principle of Galvani's experiment with frog legs (1790)
Source: Wikipedia, Wells, D.A (1859)

This was a new phenomenon: a frog's leg in a nerve-muscle preparation contracted every time the muscle and the nerve were connected by a metal arc, which usually consisted of two different metals (Figure 22). Galvani concluded from his findings that an intrinsic presence of electricity existed: the "animal electricity" (Naum Kipnis, 1987, pp. 114-116).

> *And still we could never suppose that fortune were to be so friend to us, such as to allow us to be perhaps the first in handling, as it were, the electricity concealed in nerves, in extracting it from nerves, and, in some way, in putting it under everyone's eyes* (Piccolino, 1998, p. 381).

Animal electricity had been a topic of interest for many scientists in the preceding decades who studied the relationship between electricity and life as observed in several species of fish like the electric eels (*Gymnotus electricus*).

They were looking for the "neuro-electric fluid." So news about Galvani's discovery quickly spread from Italy to France, Germany, and England. Reading about Galvani's theory as described in a fifty-three-page Latin paper, *De Viribus Electricitatis in Motu Musculari Commentarius* that was written in 1792, many scientists started repeating Galvani's experiments with frogs.

> *The news of Galvani's discovery caused great feeling in scientific circles, and it produced a repercussion among physicians and the merely curious: astonishment, wonder, and immediate resolution to repeat Galvani's experiments, with the obvious consequence that frogs were decimated in great quantities everywhere, first in Italy and then in Europe* (Bernardi, 2000a, p. 102).

An unusually large number of authors from all over Europe published their findings in the next couple of years (Naum Kipnis, 1987, p. 117). At the University of Pavia, the feeling produced by Galvani's little Latin book, which soon became a best seller, was enormous. Mariano Fontana, a member of the Imperial Chancellery of Emperor Charles V, wrote to the author, "now with endless pleasure I tell you that the result of your finest experiments is considered an original discovery, that the experiments have been repeated and found very exact...In short, here now all is animal electricity, and your name is famous in Pavia."

Nature of lightning: voltaic electricity

The work of Galvani also attracted the attention of Alessandra Volta (1745–1827), a professor of experimental physics from the University of Pavia. His work resulted in another discovery called "voltaic electricity." Volta repeated the Galvani experiments. After finding out the important role of the conductors used, he challenged Galvani's theory. In a letter he wrote in August 1796, he said, "One can consider this mutual contact of two different metals as the immediate cause that set the electric fluid in motion, instead of attributing this power to the double contact of these metals with the humid conductors..." (Naum Kipnis, 1987, p. 122).

Figure 23: Volta pile (nineteenth century)
Source: www.sparkmuseum.com

On March 20, 1800, Volta wrote a letter to Joseph Banks, the president of the Royal Society (Volta, 1800). On June 26 this letter was read before the Royal Society in London. In his letter Volta described a new source of energy: a pile of plates of silver (A) and zinc (Z) separated by cardboard (a) and soaked in salt water. The (electrochemical) battery from the spate groupings AZa, AZa, AZa, AZa, etc. was born (Figure 23). "…In this manner I continue coupling a plate of silver with one of zinc, and always in the same order, that is to say, the silver below and the zinc above it, or vice versa, according as I have begun, and interpose between each of those couples a moistened disk. I continue to form, of several of this stories, a column as high as possible without any danger of its falling."

The discussion continued and became a controversy. Scientists either adhered to Galvani's theory or to Volta's theory; it was basically the choice between the concepts of "static electricity" (Galvani's theory) and "contact electricity"—the electricity created by contact potential (Volta's theory) (Geddes & Hoff, 1971). It was a controversy that had to been seen within the context of that period in time:

> *Between 1791 and 1800, in the ten years from the publication of Galvani's Commentarius to the invention of the electric pile by Alessandro Volta, a scientific revolution occurred in Europe. It was not only a scientific controversy. The political problems and revolutionary events, which at the end of the eighteenth century changed French and Italian life, had a close relationship with the development and conclusion of the controversy between Galvani and Volta* (Bernardi, 2000b, p. 102).

These political problems and revolutionary events were related to the aftermath of the French Revolution. It was the time that Napoleon rose to power and conquered large parts of Northern Italy in his Italian Campaigns (1792–1802) and then created a vassal state. Volta, in 1801, gave a demonstration in Paris of the research that had preceded his invention. The event was attended by Napoleon I, who awarded him a gold medal. In 1805 the French emperor granted him an annuity and named him Knight of the Legion of Honor; in 1809 Napoleon made Volta a Senator of the Kingdom of Italy and, the following year, a count.

Whatever the controversy, the discovery of the "voltaic pile" would have enormous consequences on the further development of electricity.

Figure 24: Presentation of Volta's battery to Napoleon (November 1801)

Source: Bibliotheque de Napoleon. www.napoleon-livre.com

The pile was the last great discovery made with the instruments, concept, and methods of the eighteenth-century electricians. It opened up a limitless field. It was immediately applied to chemistry, notably to electrolysis, and soon brought forth the shy elements sodium and potassium from fused soda and potash. Its steady current provided the long sought means for establishing a relation between electricity and magnetism. The consequent study of electromagnetism transformed our civilization (Heilbron, 1979, p. 494).

Electricity explored

Electricity was, for a lot of curious people, an interesting phenomenon worth exploring. These early explorations had an experimental character and led to the discovery of the electric phenomena. The *experimenting scientists* of electricity were children of their time. Electricity, like heat, visible light, and magnetism, was considered to be a fluid. Franklin spoke about an "electric fluid." He propagated the single-fluid theory in contrast with the double-fluid theory of DuFray. Galvani saw the animal electricity flow from the brain, through the nerves, to the muscles: the "galvanic fluid." Volta, with his theory of metallic (voltaic) electricity, spoke about "electric fluid."

But it would take the *theoretical scientist* to really understand the nature of electricity—people such as the Dane Hans Christian Oersted (1777–1851), the Frenchman Andre-Marie Ampère (1775–1836), and the Englishmen Michael Faraday (1791–1867) and James Clerk Maxwell (1831–1879). Over time, the experimental scientist had more or less grasped at the "power of lightning" abandoning the fluid concept. Now it was the "forces" that created the basics for their thinking, such as von Leibnitz's "living force" (vis viva). In the case of electricity, these ideas were called the "field of force" concept that Faraday would use in his essay *On the conservation of Force* and that Maxwell would use in his publications *On Physical lines of Force I–IV*.

Hans Christian Oersted: electromagnetism

In 1820 Hans Christian Oersted observed, during a lecture, that a compass needle would move when an electric current passed through a nearby electric cable; it was the discovery of *electromagnetism* (Figure 25, Figure 26).

> *Oersted tried to place the wire of his galvanic battery perpendicular (at right angles) over the magnetic needle, but remarked no sensible motion. Once, after the end of his lecture, as he had used a strong galvanic battery to other experiments, he said, "Let us now once, as the battery is in activity, try to place the wire parallel with the needle," as this was made, he was quite struck with perplexity*

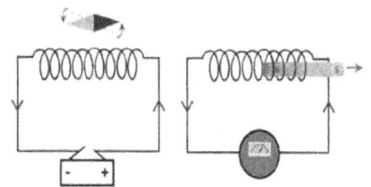

Figure 25: Principles of Oersted's electromagnetism

An electric current from a battery creates an electromagnetic field moving a compass needle (left). And a permanent magnet moved in a coil creates an electric current (right).

by seeing the needle making a great oscillation (almost at right angles with the magnetic meridian). Then he said, "Let us now invert the direction of the current," and the needle deviated in the contrary direction. Thus the great detection was made; and it has been said, not without reason, that "he tumbled over it by accident." He had not before any more idea than any other person that the force should be transversal. But as Lagrange has said of Newton in a similar occasion, "such accidents only meet persons who deserve them" (Nahum Kipnis, 2005, p. 3).

He described the phenomenon and mailed a four-page pamphlet in Latin, *Experimenta circa effectum coflictus electrici in acum magneticam* (Figure 27), to a number of renowned scientists and institutions. It created quite an interest among the "gentlemen of science" in England (Humphry Davy, William Wollaston), France (Arago, Ampère, de la Rive), Germany (Schweigger, Gilbert), Italy (Volta), and the United States.

> *Oersted's brief notice of his discovery was tested within a few weeks by some of the world's leading scientists—by Sir Humphrey Davy at the Royal Institution in London; by Dominique Arago, one of the editors of the Annales de Chimie et de Physique at the Academic des Sciences in Paris; by Auguste de la Rive, professor of chemistry at Geneva, Switzerland; by J. S. Schweigger—professor of physics and chemistry at Halle and editor of the journal Jilr Chemie und Physik; and by L. W. Gilbert, professor of physics at the university in Leipzig and editor of the Annalen der Physik und der physikalischen Chemie. All of these scientists confirmed Oersted's results* (King, 1962, p. 256).

It was François (Jean Dominique) Arago who brought news of Oersted's discoveries to Paris, after he had witnessed them during a visit to de la Rive in Geneva. The members of the Academy of Sciences were initially skeptical of his report and

Figure 26: Oersted setting up his experiment (1820)

Source: Louis Figuier: Les merveilles de la science, ou Description populaire des inventions modernes (1867), page 713.

were only convinced by his demonstration, which took place on September 11, 1819. Among those present at the demonstration was Andre-Marie Ampère.

Andre-Marie Ampère: electrodynamics

The Frenchman Andre-Marie Ampère (1775–1836) was born into a well-to-do family living alternately in the small village of Poleymieux-au-Mont-d'Or—ten kilometers from Lyon—and in the city of Lyon, a center of the silk trade in those days. His father, a prosperous silk merchant, decided not to put his son through the traditional teaching system. Instead, he led the young boy to become an autodidact by exposing him to the extensive library he had at home; he never required him to study anything, just to follow his own tastes. Ampère's early education took place in a deeply religious atmosphere. His mother was very religious and arranged for her son to be thoroughly instructed in the Catholic faith. He became interested in mathematics, but this was only one of his many interests as he acquired a considerable knowledge of metaphysics, chemistry, botany, and physics. After taking a few lessons in differential and integral calculus from a monk in Lyon, Ampère began to study works by the Swiss mathematicians Leonard Euler, Joseph-Louis Lagrange, and Jacob Bernoulli.

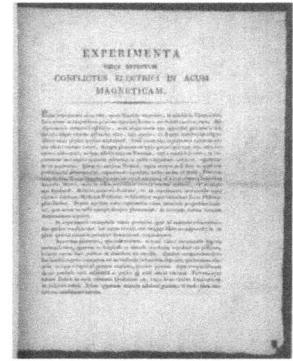

Figure 27: Oersted's *Experimenta circa effectum conflictus electrici in acum magneticam* (1820)

Source: http://www.ampere.cnrs.fr

> *France was in that time quite in turmoil, and his youth was influenced by it. The aftermath of the French Revolution caught him, when his father, a businessman who became Justice of the Peace in Lyon, was guillotined in 1793 as part of the Jacobin purges of the period. Having a close relation with his father, this was a shock for Andre, and he went into an almost mental and physical withdrawal for more than a year. In the midst of these traumatic events, Ampère met Catherine-Antoinette Carron (always referred to as Julie) who was to become his wife. Julie was somewhat older that Ampère and a member of a bourgeois family of good standing. They were married on August 7, 1799, in a clandestine religious ceremony because the revolutionary government prohibited these ceremonies* (Wisniak, 2004, pp. 166, 167).

Ampère had started teaching, and in 1802 he was appointed a professor of physics and chemistry at Bourg-en-Bresse, near Lyon. In 1803 Julie died after a difficult birth and left Ampère with a son (Hofmann, 1995). In 1804 he began a tutoring post at the *École Polytechnique* in Paris where he, in 1809, became professor of analytical mathematics. His depression (after Julie's death) contributed to his decision to take the earliest opportunity to leave Lyon for new surroundings in Paris. Later he would regret this decision. Ampère painfully missed his Lyon friends who had attempted to fill the emotional void left by Julie's death. Although Ampère gradually adjusted to the priority disputes and infighting of the Parisian scientific community, he always longed for a return to the intellectual life he had experienced in Lyon (Hofmann, 1995, p. 82).

In Paris Ampère worked on a wide variety of topics. Although a mathematics professor, his interests were broader and included—in addition to mathematics—metaphysics, physics, and chemistry. In chemistry he worked on fluorine, and he corresponded with Humphry Davy. In the course of his correspondence with Davy about fluorine, Ampère mentioned Dulong's discovery of a "detonating oil" (nitrogen trichloride) that had cost the latter an eye and a finger. Davy instantly set to work on this explosive and was himself involved in two accidents (Gardiner & Gardiner, 1965, p. 237). These experiments resulted in advances in the field of chemistry—one of them being the discovery of a substance called iodine.

During this time Ampère worked on partial differential equations, which he presented to the *Institut National des Sciences* (Academy of Sciences) in 1814. He was elected a member of the institute in the same year. This distinction earned him a professorship at the *École Polytechnique*. Ampère was present at the demonstration that Arago gave in 1819 to the members of the Academy of Sciences. This event was the starting point for Ampère's research work on electricity and magnetism, as he became excited by the discovery:

> *Depuis que j'ai entendu parler pour la première fois de la belle découverte de M. Oersted…j'y ai pensé continuellement, je n'ai fait qu'écrire une grande théorie sur ces phénomènes et tenter des expériences indiquées par cette théorie* (Gardiner & Gardiner, 1965, p. 238).[33]

Like Humphry Davy and Wollaston in England, who Ampère knew well, he started experimenting. A succession of papers resulted. He sent them to Davy where Faraday also read them. Later Faraday and Ampère, who met in 1814 during Davy's tour through Europe, would correspond

---

[33] Translation by author: "Since I first heard of the beautiful discovery of Mr. Oersted…I thought about it all the time; I only wrote a large theory on these phenomena and experimented based on this theory."

intensively about their findings through experimentation (Ross, 1965). He also complained to Faraday of the activities of his enemies, who cast doubts on his experiments and sought to harm his reputation abroad. At this time (1825) Faraday had every reason to sympathize, as the rift between Davy and himself had just become apparent. In a letter he wrote:

> *"I am sorry to find by one of your letters that you experience an unworthy opposition to the fair and high claim you have to the approbation and thanks of your fellow Philosophers. This, however, you can hardly wonder at. I do not know what it is nor by whom exerted in your case, but I never yet even in my short time knew a man to do anything eminent or become worthy of distinction without becoming at the same time, obnoxious to the cavils and rude encounters of envious men. Little as I have done, I have experienced it and that too where I least expected it."* (Gardiner & Gardiner, 1965, p. 243)

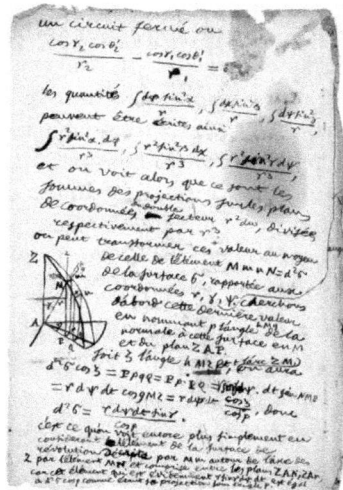

Figure 28: Extract from Ampere's calculations concerning the interactions between circuits
Source: http://www.ampere.cnrs.fr

In 1820, after repeating Oersted's experiment, Ampère started experimenting with electricity. He realized that the magnetism could be enhanced if a wire was wound into a helix or coil and an iron needle placed in the center of it. On September 25, 1820, he showed that a wire coiled in a spiral acted just like a magnet. A few weeks later, he discovered that two rectilinear, current-carrying wires attracted and repelled each other according to the directions of the current.

> *Suivant le sens dans lequel on fait passer le courant dans une tell spirale, elle est en effet fortement attirée ou repoussée par le pole d'un aimant qu'on lui présente de manière que la direction de son axe sot perpendiculaire au plan de la spirale…En replaçant l'aimant par une autre spirale dont le courant soit dans le même sens que le sien, on a de même attractions et répulsions* (Ampère, 1821, p. 60). [34]

---

[34] Translation by author: "According to the sense in which the current in a such a spiral is passed, it is indeed strongly attracted or repelled by the pole of the magnet that is presented in such a way that the direction of its axis is perpendicular to the plane of the spiral…By

These experiments (Figure 29) showed Ampère that two parallel wires attracted each other when they were carrying currents flowing in the same direction, and they repelled each other if the currents ran in opposite directions.

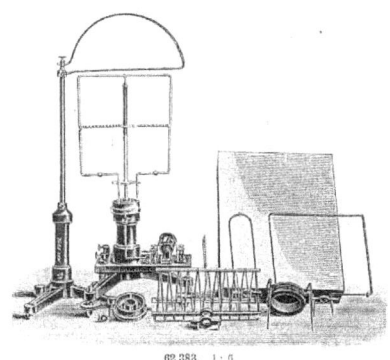

He maintained that it was the same for two current elements that were infinitely small and parallel. Furthermore, Ampère was led to suppose that the force between two elements was zero, if one of them were situated in a plane perpendicular to the other element in its middle. He then arrived at an expression for the elementary force being proportional to: g h (sina sinβ cosγ+ k cosa cosβ) / r², where g and h depended on "the electricity passing in equal time periods." Here is a first definition of the notion of the intensity of a current (C. W. Blondel, B., 2012).

Figure 29: Ampère's Stand: instruments Ampère used to experiment with the relation between electricity and magnetism
Source: Max Kohl Catalog. www.evm.edu

The discovery of the attractions and repulsions of rectilinear, current-carrying wires marks the end of Ampère's early discoveries in electrodynamics (Williams, 1983, p. 507).

After recovery from another illness, it was a discovery by Faraday (Figure 34) that relaunched Ampère's research in the autumn of 1821. Faraday announced that he had achieved the continuous rotation of a magnet under the action of a conductor and vice versa. These continuous rotations astonished Ampère, and he started experimenting and publishing again. He wrote to a friend:

Depuis que le mémoire de M. Faraday a paru je ne rêve plus que courants électriques. Ce mémoire contient des faits électromagnétiques très singuliers qui confirment parfaitement ma théorie quoique l'auteur cherche a la combattre pour lui en substituer une de son invention [35]
(Gardiner & Gardiner, 1965, p. 240).

---

replacing the magnet by another spiral, whose current is in the same direction as his own, one has the same attractions and repulsions."
[35] Translation by author: "But since Faraday's memoir has been published, I dream only of electrical currents. This memoir contains some very unusual facts about electromagnetism,

In 1821 he published *Mémoire sur l'action mutuelle entre deux courants électriques, un courant électrique et un aimant ou le globe terrestre, et entre deux aimants*. There he outlined the basics for the nature of electricity:

> *L'action électromotrice se manifeste par deux sortes d'effets...J'appellerai le premier tension électrique, le second courant électrique. Le premier s'observe lorsque les corps entre lesquels l'action électromotrice a lieu sont séparés l'un de l'autre par des corps non conducteurs dans tous les points de leur surface autre que ceux où elle étais établie: le second est celui où ils font, au contraire, partie d'une circuit de corps conducteurs qui les font communiquer par des points de leur surface différents de ceux ou se produit l'action électromotrice*
> (Ampère, 1821, p. 3) [36]

This publication was followed in 1822 with *Recueil d'observations electro-dynamiques* and *Exposé des nouvelles découvertes sur l'électricité et le magnétisme de MM. Oersted, Arago, Ampère, Davy, Biot, Erman, Schweiger, De La Rive, etc*. In 1826 he published *Description d'un appareil électro-dynamique* (A.-M. Ampère, 1826) and *Théorie des phénomènes électro-dynamiques: uniquement déduite de l'expérience* (A. M. Ampère, 1826). In this paper Ampère gave a new name to the phenomena he studied and explained—*electrodynamics*—and brought to a close his feverish work over the previous years on the new science of electrodynamics.

Oersted's discovery of electromagnetism and Ampère's theory of electrodynamics resulted in a frenzy of scientific activity. In London, in the laboratory of the Royal Institution, Sir Humphry Davy originated studies on electromagnetism in the first flush of interest and enthusiasm with which he had greeted the news of Oersted's discovery. Ampère's rapid development of the subject gave fresh food for thought, particularly the theory that magnetism could be explained by postulating electrical currents within each atom—an idea that sounds marvelously similar to our modern knowledge of atomic structure (Ross, 1965, p. 197).

Michael Faraday, a young assistant that Davy had hired after an accident when chemical experiments damaged his eyesight, was involved in all the work. Andre-Marie Ampère, Humphry Davy, and Faraday thus became

---

which perfectly confirm my theory, although the author tries to dispute it by substituting one of his own invention."

[36] Translation: "Electromotive action manifests itself in two kinds of effects...I will call the first electrical voltage, the second electric current. The first occurs when the bodies between which the electromotive action takes place are separated from the other by nonconductive bodies in all points of their surface other than where it was established; the second is the one where they are, on the contrary, part of a circuit of conductive bodies that make them communicate through points of their surface different from those where the electromotive force action is produced."

linked in their scientific efforts—with some friction, as Davy forgot to mention Ampère's work, Ampère did not mention Davy at the right moment, and Faraday was too hasty and forgot to mention Davy's and Wollaston's work (Gardiner & Gardiner, 1965). Later, Ampère was honored in England when he was asked to become, in 1827, a (foreign) Member of the Royal Society of London.

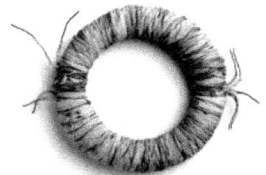

Figure 30: Faraday's induction ring, replica (1831)
Source: Science Museum Group, Collections Online, Objects

James Clerk Maxwell, in his *Treatise on Electricity and Magnetism*, later wrote of Ampère and his achievement:

> The experimental investigation by which Ampère established the laws of the mechanical action between electric currents is one of the most brilliant achievements in science. The whole, theory and experiment, seems as if it had leaped, full grown and full armed, from the brain of the "Newton of Electricity" (Gardiner & Gardiner, 1965, p. 245).

Ampère explained the mechanism of electricity in his general theory connecting electric currents with magnetic forces (Steinle, 2002, pp. 414-415). So he explained the mechanism behind Oersted's discovery where an electric current influenced the magnetic needle. But his theory did not explain the reverse action: magnetism influencing electric current. If the presence of an electric current is always concomitant with a magnetic field, why should it not be possible to reverse Oersted's experiment and induce electric currents by the action of a magnet? (Ross, 1965, p. 184)

Michael Faraday, who was also intrigued with Oersted's discovery, provided the explanation. He studied the question and experimented in 1831 with a soft iron ring with two sets of coils (as seen more or less in today's transformer) (Figure 30). Connecting a battery to the first coil resulted in current in the second coil. He had found the induction effect and thus expanded the relation between magnetism and electricity: the *electromagnetic induction* (Steinle, 2002, pp. 416-417). This was the creation of a "potential difference" when a conductor is exposed to a varying

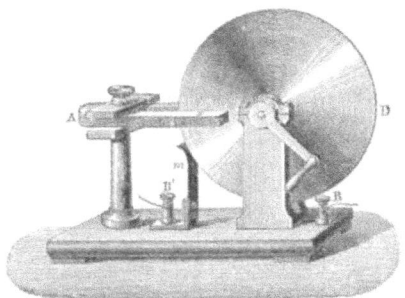

Figure 31: Faraday's wheel: a generator of electricity (1831)
Source: Wikimedia Commons

magnetic field. He created the "Faraday wheel," a generator of electricity (Figure 31). Faraday's "law of induction" explained the interaction between an electric circuit and a magnetic field—the basic operation principle for electric motors, solenoids, and generators. Interestingly, Faraday's open publication of his discoveries without applying for a patent created a situation where others could only patent "improvements" to his ideas (Arapostathis & Gooday, 2013, p. 114).

The Frenchman Francois Arago (1786–1853), educated at the *Ecole Polytechnique* in Paris and secretary of the Academy of Sciences, discovered the effect to be known as Arago's rotations. In 1824 he demonstrated that a rotating copper disk produced rotation in a magnetic needle suspended above it (Figure 32).

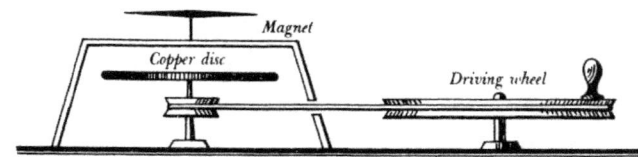

Figure 32: The Arago rotation experiment by which a magnetic needle is made to drag after a revolving copper disk (1825) (side view)
Source: (Ross 1965, p.193)

Scientists such as Charles Babbage and Herschel repeated his experiments (Babbage & Herschel, 1825). And the Frenchman J. D. Colladon (1802–1893), with his experiments in 1825, almost discovered electromagnetic induction (Ross, 1965, pp. 192-193).

Michael Faraday later proved these to be induction phenomena. It was then William Sturgeon (1783–1850) who in 1825 conceptualized that electricity and the properties of metal could create a magnetic force; the electromagnet was born. Finally it was James Maxwell who in 1861, by creating his theory of classical electromagnetism known as the "Maxwell Equations," demonstrated that electricity, magnetism, and light are all manifestations of the same phenomenon, the electromagnetic field.

So the scientists observed the phenomenon of magnetic induction (Oersted, Arago) and were able to explain the principle behind it (Faraday, Ampère). Then the "electricians" applied it (Sturgeon, Henry) and translated it into manageable entities (for example products such as electric lamps and motors). And James Clerk Maxwell explained the mathematics by combining magnetism and electricity in one theory.

Humphry Davy

Humphry Davy (1778–1829) was one of the first professional scientists, earning his living and rising spectacularly from an impoverished upbringing in Cornwall to be president of the Royal Society and a baronet. He owed his rise to patronage as well as to his range of abilities: as a lecturer, as a chemical theorist, and as a very early applied scientist. His exalted position brought him little happiness, for he could not satisfy all the hopes put upon him as the successor to Sir Joseph Banks, who was president of the Royal Society for over forty-one years (1778–1820). Admired rather than loved, he became unpopular and was seen as haughty. In his last two years, spent wandering lonely and ill in Italy and the Alps, he sought to make sense of his life, writing dialogues as his bequest to the new generation (Knight, 2000).

> *Davy too had intellectual powers amounting to genius; he was, moreover, a romantic idealist who foresaw great practical outcomes from science for the benefit of mankind. This was a new viewpoint for the man of science, and a stimulus for investigation more powerful than any other. He was always eager, therefore, to move from the realm of theory to that of practice* (Ross, 1965, p. 197).

When Davy learned in 1820 about Volta's discovery of chemically produced electricity—the voltaic battery—he became interested, and it would lead to his fame in electrochemistry. His interest in electricity was quite understandable, as he had already experimented for a long time with elementary chemistry in relation to the "voltaic electricity." As early as 1806, he gave a lecture on the chemical effects produced by electricity and water, the decomposition of various compounds, and the transfer of "constituent Parts of Bodies by the Action of Electricity" (the Bakerian Lecture: *On Some Chemical Agencies of Electricity*). Here he concluded about the importance of electricity:

> *Alterations of electrical equilibrium are continually taking place in nature; and it is probable that this influence, in its faculties of decomposition and transference, considerably interferes with the chemical alterations occurring in different parts of our system. The electrical appearances which precede earthquakes and volcanic eruptions, and which have been described by the greater number of observers of these awful events, admit of very easy explanation on the principles that have been stated. Besides the cases of sudden and violent change, there must be constant and tranquil alterations in which electricity is concerned, produced in various parts of the interior strata of our globe* (J. Davy, 1839).

Davy went on a lecture tour between 1813 and 1815, in a Europe troubled with conflicts, and to collect a medal that Napoleon and the Institute de France had awarded him for his electrochemical work.[37] Faraday accompanied Davy as a personal servant on a long trip through Europe (planning to visit France, Switzerland, Germany, Italy, Greece, and Turkey) (Figure 33). It gave Faraday the opportunity to get to know the scientific elite of Europe (such as de la Rive, Volta, Ampère, Arago, Gay Lussac) and to visit the centers of science of those days (Geneva, Paris, Florence). About these visit Dumas wrote:

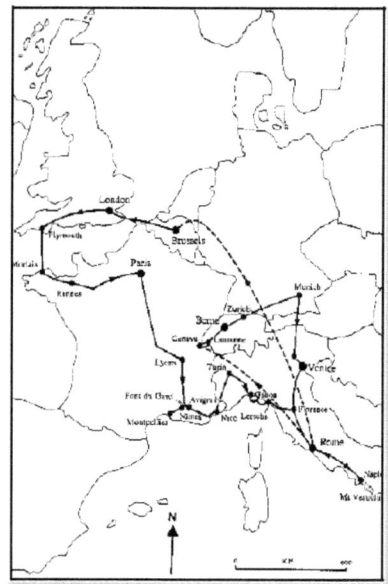

Figure 33: Humphry Davy's "Grand Tour" to Europe 1813–1815
Source: (Bowers & Symons, 2006, p. 60)

> *His laboratory assistant, long before he had won his great celebrity by his works, had by his modesty, his amiability, and his intelligence, gained most devoted friends at Paris, at Geneva, at Montpellier. Amongst these may be named in the front rank M. de la Rive, the distinguished chemist, father of the illustrious physicist whom we count amongst our foreign associates. The kindnesses with which he covered my youth contributed not a little to unite us— Faraday and myself. With pleasure we used to recall that we made one another's acquaintance under the auspices of that affectionate and helpful philosopher whose example so truly witnessed that science does not dry up the heart's blood. At Montpellier, beside the hospitable hearth of Berard, the associate of Chaptal, doyen of our corresponding members, Faraday has left memories equally charged with an undying sympathy which his master could never have inspired. We admired Davy; we loved Faraday* (Thompson, 1898, p. 20).

---

[37] The given rationale of making contact with European scientists and collecting the medal from the "Institute Imperial" was only for public consumption. Davy and Lady Jane were part of the wealthy British upper class, and they were on a Grand Tour—a mix of holiday and dutiful obligation all members of their set did at least once. While on the continent, they indulged the uniquely English feeling of moral superiority to those unfortunate enough to live in other countries, and collected mild adventures they would dine out on for months after their return (Bowers & Symons, 2006, p. 78).

The Faraday memoirs about this trip to Europe give insight in the relation between Faraday and Davy. The first is what would be called later the "valet incident" during the trip to Europe. When Davy asked Faraday to accompany him on his travel in Europe, Faraday had not been more than a couple of miles from London in his young life.

> *[On the trip] There were to be five in the party—Faraday, Lady Jane, her maid Mrs. Meek…Davy, and his valet, La Fontaine. However, at the last minute that henpecked servant "was diverted by the tears of his wife" and refused to go. Faraday told friends and relatives he was engaged as Davy's "philosophical assistant" and "amanuensis," "assisting in experiments in taking care of the apparatus and of his papers & books and in writing and other things of this kind."…On 13 October Sir Humphry, Lady Jane, with Mrs. Meek sitting beside her and twenty-two-year-old Michael Faraday relegated to the boot seat atop the Davys' private carriage, bounced along Park Lane swaying in tune with the coach's gentle rhythms as they headed south toward the port city of Plymouth* (Bowers & Symons, 2006, p. 56).

So it was Faraday, sitting on top of the carriage like a servant, who would be acting as valet. It was understood that the valet activities would be temporary till the moment Davy could hire another valet sometime along their voyage.

> *They reached Plymouth on 15 October where they stopped at a commercial inn. The accommodations were comfortable, and Faraday looked forward to a restful night's sleep, but he had his newly acquired valet duties to perform before retiring—he must fetch the water necessary for shaving Sir Humphry, turn down his bed, pack used clothing then lay out a clean outfit for the next day, refill the water pitcher, and, most educational of all, empty the chamber pot. It has been said no man is a hero to his valet, and Faraday's perspective of Humphry Davy was revised quickly. Two days on the road as sightseer and valet left him exhausted, but that night he wrote in his journal of his joy at being introduced to the pleasures of travel and his eager anticipation of things to come* (Bowers & Symons, 2006, p. 61).

But there were more trials to come as Davy's wife, Lady Jane, (a twenty-seven-year-old, wealthy widow from Shuckburgh Apreece, first baronet of Washingley, Huntingdonshire, who married Davy in 1812) proved to be a decisive factor during the voyage. During this period the worst traits of her character showed themselves.

> *Davy may have considered Faraday a professional colleague, but his wife thought him little more than a slave…In Davy's mind his marriage into the upper levels of London society merely conferred on him what was his natural right of superiority. Lady Jane, polite and correct in society, gave off the condescending airs of the aristocracy in her presumption, conceit, and volatile irascibility when dealing with the help…*

> *Faraday never gave any details about his clashes with Lady Jane but others did. While in Geneva the travelers met up with Dr. Alexander Marcet and his wife Jane, the author of Conversations in Chemistry, one book that so intrigued Faraday when he read it in Riebau's shop. The Marcets and some friends gave a small dinner party welcoming Davy, Lady Jane, and Faraday to the city. When Jane Marcet announced dinner and was about to lead her guests into the dining room, Lady Jane held Faraday back, saying "Mr. Faraday, you will now go and eat your meal in the kitchen." The humiliated young man was crestfallen and had no choice but to go below stairs and share his meal with servants as the others took their places at the table. When the meal was finished and the ladies rose to leave the men to their port, Dr. Marcet remarked in a loud whisper "and now, my dear Sirs, let us go and join Mr. Faraday in the kitchen"* (Bowers & Symons, 2006, pp. 57,58).

As the political situation in France was deteriorating rapidly after Napoleon's disastrous Russian Campaign in 1812, they travelled from Paris to Montpellier (Figure 33), a trip of nine days. From there they travelled to Aix en Provence, Nice, and, passing the Alps, to Turin. The passing of the snow-covered Col de Tende (1870), with the help of sixty-five hired people, was quite an adventure.[38] Faraday described it in his diary as follows (Jones & Faraday, 2010):

> *Saturday, 19th. [February 1870]—Col de Tende. Rose this morning at daybreak, which was much advanced at half-past five o'clock, and made preparations for crossing the great mountain, or Col de Tende. At Tende the noble road, which had given such facile and ready conveyance, finished, and it was necessary to prepare for another sort of travelling. Expecting it would be very cold, I added to my ordinary clothing an extra waistcoat, two pairs of stockings, and a nightcap: these, with a pair of very strong, thick shoes and leathern overalls, I supposed would be sufficient to keep me warm…*

---

[38] Today one crosses de Col de Tende (using a tunnel to avoid the actual pass) in about 4.5 hours—a distance of 364 km between the villages of Tende and Limone Piemonte.

The road began to change soon after leaving Tende, and at last became nothing but ice. It was now fit for beasts of burden only: grooves had been formed in it at equal distances to receive the feet of the horses or mules, and prevent their falling; and though convenient to them, it was to us a great evil, for as the wheels fell successively into the ruts, it produced a motion not only disagreeable, but very dangerous to the carriage…and in about half an hour afterwards we came to a halt, and the end of the carriage road. Here on an open space the rest of the men who were to conduct and convey us and the baggage over the mountain were collected, and the scene was a very pretty subject for the pencil…

The horses being taken off, all hands worked to dismount the carriage and charge the traineaux, and after some time this was done. The pieces of the carriage were placed on two sledges, and the rest, as the wheels, boxes, &c., loaded five mules…The traineaux with the body of the carriage had started about twelve o'clock. After they had been loaded, ropes were fixed to them at different parts, and they were consigned each traineau to about twenty men, who were by main strength to haul it over the mountain. They set off with a run and loud huzzas; but the mules were not ready until one o'clock, and as a mule driver could be better spared, if wanted, than a man from the sledges, I kept in their company…

After some climbing and scrambling, the exertion of which was sufficient to keep me very comfortably warm, I reached a ruined, desolate house, half-way up the mountain. Here we found the traineaux; the men, having rested themselves after this long and laborious stage, were now waiting for their leader and the dram bottle…About half-past four we passed a little village consisting of seven or eight huts nearly buried in the snow; they were uninhabited, and are principally intended as a refuge for the men if accidents or other circumstances should occur in the mountains during the night. At about a quarter past five, evening began to come on, and the effect produced by it on the landscape was very singular, for the clouds and the mountains were so blended together that it was impossible to distinguish the earth from the atmosphere…Just as the starlight came on, the sounds of the evening bell of a distant village were faintly heard…we got to Leman about seven o'clock in the evening, and there put up for the night; supper and rest being both welcome.

On February 22 they finally arrived in Turin, finding the city's Carnivale festivities in full swing. After Turin they went to Genua, visiting the opera there. Sir Humphry made contact with Professor Viviani, a local chemist who had several electric fish in captivity, and with Faraday he took advantage of the opportunity to make some electrical experiments with them. They determined that the intensity of the electric currents the fish produced was extremely small and unable to trigger the electrolysis of water. From Genua they travelled by boat on the rough Mediterranean seas to Sestri Levante[39] (Bowers & Symons, 2006, p. 80). After visiting Florence and Rome, they went as far as Naples and visited the Vesuvius. There they collected minerals.

> *Of particular interest to Davy was the large deposit of iron chloride they discovered on the rim. They investigated it closely and even took samples, then suddenly the wind changed and everyone had to bid a hasty retreat from the poisonous cloud threatening to engulf them* (Bowers & Symons, 2006, p. 89).

From there they went up north, crossing the Alps for a second time, visiting Geneva (Switzerland), Munich (Germany), and returning by Venice and Florence to Rome (Italy) for a second time. However, the planned continuation to Turkey was cancelled after Napoleon escaped from Elba on February 26, 1815.

> *Suddenly movements between countries were restricted, and the route to Constantinople was quarantined. Davy's nationalism, as well as his hatred of Napoleon, was aroused, so he decided to return to England immediately. Because travel through France was out of the question, they had to take a route through Austria, Germany, Holland, Belgium, then across the North Sea to England* (Bowers & Symons, 2006, p. 95).

In April 1815 they were back in England, after an exciting voyage of thirty months on the European continent in the last days of Napoleon's rule.

The chemist and physicist Davy (knighted in 1812, created a baronet in 1818, and becoming President of the Royal Society in 1820 after Sir Joseph Banks), was fascinated by Volta's discovery of the voltaic pile (which was by its nature an electrochemical device). As were so many other scientists, he and his colleague William Hyde Wollaston were intrigued by Oersted's discovery and experimented with the phenomenon of electricity and magnetism. The very day that Oersted's memoir was published in England, Davy took a copy down into the laboratory of the Royal Institution, and he and Faraday at once set to work to repeat the experiments and verify the facts (Thompson, 1898, p. 80). Davy wrote Wollaston a detailed account on his findings, which he concluded with:

---

[39] Sestri Levante is located on the Italian coast, halfway between Genua and la Spezia. Today the overland travel of some fifty kilometers on the highway takes about forty minutes.

> *The experiments detailed in these pages were made with the apparatus belonging to the Royal and London Institution; and I was assisted in many of them by Mr. PEPYs, Mr. ALLEN, and Mr. STODART, and in all of them by Mr. FARADAY* (H. Davy, 1821, p. 18).

The concept of electricity being related to magnetism, was also a hot topic in Davy's laboratory.

> *At first, Davy was occupied by Wollaston's idea that the newly discovered effect might be used to produce rotatory motion. When their initial attempts to bring this about had failed, other implications of the discovery remained to be explored…The possibility of producing an electric current by means of magnetism appeared to Davy to be a direct outcome of Ampere's theory of magnetism, and it seems that Davy himself made some unsuccessful efforts to realizing it* (Ross, 1965, p. 197).

Michael Faraday: electromagnetic induction

The discovery of electromagnetic induction sounds simple—a cascading development of ideas, concepts, and theories by a multitude of scientists—but in reality it was a struggle in which a lot of nontechnical aspects that dominated the scene. A struggle with that was related to the personalities involved; their background and uprising, their characters, ambitions, hopes and fears. But also the interpersonal relations in which the interaction took place; the scientific competition among peers, the honor of being the first to invent, the unmistaken jealousy and envy. For those who were comfortably of, the gentlemen of science with their wealthy upbringings, ample financial means to finance their livelihood and their experimenting, the persuit of science was a way of spending one's time. For those experimental physicists with a more modest, or even impoverished background, often surviving in a life of hardship, the art of science asked many sacrifices, dedication and perseverance. But they had one thing in common: curiosity about what took place in the natural world around them, and the intelligence to observe and interpret.

Take the example of the life of Faraday, considered to be one of the great contributors to the science of electricity. Michael Faraday (1791–1867) was born in the London region as the son of a poor blacksmith in a religious family. He received just a basic education. Faraday was apprenticed as a bookbinder for seven years to bookbinder and stationer George Riebau, which gave him the opportunity to read much and educate himself. It was at the age of twenty that he attended the first of a series of four lectures on the science of chemistry by Humphry Davy at the Royal Institution in London. The subject of Davy's lecture was radiant matter, and Faraday came prepared to learn. He took detailed notes, and, since he was a bookbinder by profession, he made a three-hundred-page book out of his notes that he presented to Davy. He became, after Davy had damaged his eyesight in a laboratory experiment, Davy's temporary assistant. Three months later, on Davy's recommendation, Faraday became a chemical assistant, in 1813 at the Royal Institution. This association with Humphry Davy would become very important in the life of Michael Faraday.

Faraday originally started assisting Davy with work in the field of chemistry and then expanded into electrochemistry (resulting in his later Law of Electrolysis). His first understanding of electricity came when he wrote, on request of his friend Richard Phillips, the "Historical Sketch of Elektromechanism" for the *Annals of Science* (Faraday, 1821). But it was this analysis of Oersted's discovery (of the moving compass needle related to an electrical current) that helped him become interested in the subject of electromotive rotation. Why was it that the electric current made an "effort" to move the compass needle?

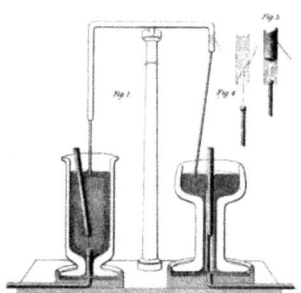

Figure 34: Faraday's experiment demonstration of electromagnetic rotation
Source: WikiMedia Commons

> *To translate this "effort" into an actual rotation, Faraday devised a most ingenious and simple apparatus. A magnet was stuck upright in a piece of wax at the bottom of a deep basin, and then the basin was filled with mercury until only the pole of the magnet was above its surface. A wire, free to revolve around the magnetic pole, was connected to a galvanic circuit. When the current was turned on, the wire rotated around the magnet. In a similar fashion, Faraday arranged things so that the magnet would rotate around the wire. The first electric motor had been invented; the rotatory power of the magnetic force surrounding a current-carrying wire was made obtrusively manifest; the conversion of electricity into mechanical work had been achieved, lending still further weight to [Faraday's] belief in the convertibility of all natural forces* (Williams, 1965, pp. 156-157).

Where Davy and Wollaston had failed, Faraday succeeded, creating the homopolar motor in 1821 (Figure 34). He had realized that the magnetic field of an electric current and the field of a magnet were always perpendicular. It was this fundamental phenomenon that created the rotative motion in Oersted's compass needle.

> *In September 1821, Faraday discovered how to produce electromagnetic rotations, bringing to a successful conclusion Wollaston's thought that such a motion might be possible. With the publication of this result, Faraday moved at one bound into the forefront of those actively engaged in developing the new science of electromagnetism* (Ross, 1965, p. 200).

He published his findings of the research on the electromagnetic rotations in an article "On some new Electromechanical Motions, and on the Theory of Magnetism" in the *Quarterly Journal of Science* of October, Volume 12, 1821 (and reprinted in the second volume of the *Experimental Researches in Electricity*. This article was the occasion of a very serious misunderstanding between Faraday and Davy and Dr. Wollaston and his friends (Thompson, 1898, p. 98). In his eagerness to publish, Faraday forgot to mention the work and ideas of Davy and Wollaston, and he was more or less accused of plagiarism.

> *That Faraday, however, should simply have had the idea directly from Davy, and that it should have come to Davy as a result of reading Ampère's papers, is extremely probable...Faraday was caught up time and again in the sweep of Davy's activities. This occasion differs only from the others in being the most illustrious. It does not detract from Faraday's great merits to do justice to his patron and teacher, whose brilliance at grasping the wider implications of phenomena was precisely his strongest trait* (Ross, 1965, p. 198).

The mere fact that Faraday did not recognize Davy's and Wollaston's contributions to the development of his novel ideas about electricity, magnetism, and rotation, would have quite some consequences later in time, in regards to his election as a Fellow of the Royal Society, his relationship with Davy, and his later work.

Faraday's rise to notoriety within academic circles continued in 1821 when he was appointed the "Acting Superintendent" of the Royal Institution, responsible for the day-to-day running of the house and its servants. Faraday had become, after his marriage in 1821, a member of the Sandemanian Church—a small sect that broke away from the Church of Scotland—where he served as a deacon and elder.

> *Faraday was appointed to the Deacon's office in 1832 and to the Elder's in 1840, appointments reflecting his high moral standing in the Sandemanian community. However, on 31 March 1844 he fell from grace and was not only removed from the Elder's office but excluded from the sect that had been his spiritual sanctuary...*

Faraday's exclusion on 31 March 1844 appears to have resulted from a searing dispute over discipline that reverberated throughout the Sandemanian churches and not from the lapse of an individual who, supposedly, visited the Queen one Sabbath and was then insufficiently penitent (Cantor, 1989, pp. 433, 437).

What really cemented his popularity within all levels of society was that Faraday's religious beliefs compelled him to decline excessive wealth and demanded a level of humility from him. This lack of interest in public standing and lack of interest in climbing the social ladder is what arguably most endeared him to the wider public.

In 1824 Faraday was elected as an official member of the Royal Society and was appointed director of the laboratory in the Royal Institution in 1825. It was Faraday's candidacy for fellowship of the Royal Society that created a second controversy. This candidacy, proposed by his friend Philips, was done without consulting Davy (who was President of the Royal Society), but he was supported by many of the existing fellows. Davy opposed Faraday's election, but nevertheless Faraday was elected on January 8, 1824. Faraday's relationship with Davy would never be the same as before.

This row also revived the controversy over the accusations made in October 1821 concerning Faraday's originality in the discovery of electromagnetic rotations. Davy repeated, apparently in error, the accusation that Faraday had used the work of Wollaston on electromagnetism without due acknowledgment…Thus ended Davy's and Faraday's personal relationship as Faraday noted in 1835 (Faraday & James, 1991, p. xxxiv).

Whenever I have ventured to follow in the path which Sir Humphry Davy has trod, I have done so with respect and with the highest admiration of his talents, and nothing gave me more pleasure in relation to my last published paper, the Eighth Series than the thought that whilst I was helping to elucidate a still obscure branch of science, I was able to support the views advanced twenty-eight years ago, and for the first time, by our great philosopher…

I have such extreme dislike to controversy that I shall not prolong these remarks, and regret much that I have been obliged to make them. I am not conscious of having been unjust to Sir Humphry Davy, to whom I am anxious to give all due honour (Faraday, 1835, pp. 341-342).

It is clear that Faraday and Davy had had an intense relationship. It was more than an apprenticeship:

> *There seems no good reason to suppose that Davy was jealous of Faraday, whose achievements at this stage were not very noteworthy—but he was beginning to fear the loss of creative scientific imagination, his marriage was unsatisfactory, and managing the Royal Society was frustrating. Faraday, at the same time, was entering into what was to be a very happy marriage and committing himself to full membership of the Sandemanian Church. Although Davy and Faraday were two of the greatest scientific orators, they were like father and son, unable to express their feelings to one another without becoming distant and formal. It was a sad breach* (Knight, 2000, p. 168).

Faraday went on experimenting, but it was not until after Davy's death in 1829 that he continued his electrical experiments. He became, in 1829, a member of the Resident Scientific Committee that gave advice to the Navy Board. He also became professor of chemistry at the Royal Military Academy of Woolworth. These electrical experiments led in 1831 to the discovery of "mutual induction": today's electromagnetic transformer. He wrote about his motivation to start experimenting in this field:

> *These considerations [the work of Ampère, Arago], with their consequence, the hope of obtaining electricity from ordinary magnetism, have stimulated me at various times to investigate experimentally the inductive effect of electric currents. I lately arrived at positive results; and not only had my hopes fulfilled, but obtained a key which appeared to me to, open out a full explanation of ARAGO'S magnetic phenomena, and also to discover a new state, which may probably have great influence in some of the most important effects of electric currents* (Faraday, 1832, p. 126).

He described his experiments in great detail.

> *A welded ring was made of soft round bar-iron, the metal being seven-eighths of an inch in thickness, and the ring six inches in external diameter. Three helices were put round one part of this ring, each containing about twenty-four feet of copper wire one-twentieth of an inch thick; they were insulated from the iron and each other, and superposed in the manner before described, occupying about nine inches in length upon the ring* (Faraday, 1832, p. 131).

Faraday concluded:

> *The various experiments of this section prove, I think, most completely the production of electricity from ordinary magnetism. That its intensity should be very*

> *feeble and quantity small, cannot be considered wonderful, when it is remembered that like thermo-electricity it is evolved entirely within the substance of metals retaining all their conducting power. But an agent which is conducted along metallic wires in the manner described; which, whilst so passing possesses the peculiar magnetic actions and force of a current of electricity; which can agitate and convulse the limbs of a frog; and which, finally, can produce a spark by its discharge through charcoal, can only be electricity…The similarity of action, almost amounting to identity, between common magnets and either electro-magnets or volta-electric currents, is strikingly in accordance with and confirmatory of M. AMPÈRE'S theory, and furnishes powerful reasons for believing that the action is the same in both cases; but, as a distinction in language is still necessary, I propose to call the agency thus exerted by ordinary magnets, magneto-electric or magneto-electric induction* (Faraday, 1832, pp. 138-139).

Through hundreds of experiments, Faraday showed that the electromagnetic effects could be explained pictorially, using lines of force that fill the space around charges and currents. This was a new paradigm in physics—the force field—that would most strongly influence Maxwell.

James Clerk Maxwell: electromagnetism in mathematical terms

James Clerk Maxwell (1831–1879) was a descendant of a very old Scotch family—the Clerks of Penicuik, near Edinburgh—and was connected by family ties to the Maxwells of Middlebie, in Dumfriesshire. He was raised in rural Scotland on the family estate of Glenair and went to the Edinburgh Academy, one of the best schools in Scotland. Next he attended the University of Edinburgh at the age of sixteen, and in 1850 he transferred to Trinity College, Cambridge University, in England. He then received a fellowship to Cambridge, became a tutor himself for a while, and then, in 1856, became a professor at Marischal College in Aberdeen, Scotland. Next, Maxwell became a professor at King's College, London, and remained there for six years. It was at King's College that Maxwell performed his most important research.

> *Maxwell first met Faraday in 1860, shortly after he assumed his place as professor at King's College. His contact with Faraday at the Royal Institution, where Maxwell lectured in 1861, made him an admirer not only of Faraday the man, but also of Faraday the experimenter. Maxwell was engaged, in particular, by Faraday's concept of the nature of the space or field existing around a magnetized or electrified body…In applying his analytic mind and mathematical command to electromagnetic problems, Maxwell did not follow the French school (Coulomb, Laplace, Poisson, and Ampère), which regarded electrical and*

magnetic phenomena as instances of action at a distance. Faraday's experiments in giving reality and form to magnetic and electrostatic fields with their lines of force emanating from a magnetic pole or a charged electric point prompted Maxwell to examine the physical properties of the surrounding space (Dibner, 1964).

In a letter dated February 20, 1854, Maxwell asked his fellow student William Thomson for advice in studying the new science of electricity:

> *If [one] wished to read Ampère, Faraday, &c how should they be arranged, and at what stage & in what order might he read your articles in the Cambridge Journal? If you have in your mind any answer to the above questions, three of us here would be content to look upon an embodiment of it in writing as advice* (Larmor, 1937, p. 3; James Clerk Maxwell, 1990).

Thompson, at that time professor of mathematics at the University of Glasgow, was well informed about Faraday's work, having written several papers based on it. So he shared with Maxwell the challenge presented by interpreting Faraday's written experimental results using mathematical formalism.

Maxwell began his research by reading Thomson's papers on the subject. In 1856 he published *On Faraday's lines of force*. It was read in two parts to the Royal Society, London, on December 10, 1855, and February 11, 1856 (J. Clerk Maxwell, 1864). The paper translated some of Faraday's ideas into mathematical language. Continuing his interest in electricity and magnetism, Maxwell wrote in the years 1861–1862 a four-part paper called *On Physical lines of Force I–IV*. In the introduction he stated:

> *I propose now to examine magnetic phenomena from a mechanical point of view, and to determine what tensions in, or motions of, a medium are capable of producing the mechanical phenomena observed. If, by the same hypothesis, we can connect the phenomena of magnetic attraction with electromagnetic phenomena and with those of induced currents, we shall have found a theory which, if not true, can only be proved to be erroneous by experiments which will greatly enlarge our knowledge of this part of physics* (J. Clerk Maxwell, 1861).

His explanations (Figure 35), unreadable for someone not educated in mathematics, resulted in the equations of electromagnetism in conjunction with a "sea" of "molecular vortices," which he used to model Faraday's lines of force. Then, in 1865 he published *A Dynamical Theory of the electromagnetic field*:

The theory I propose may therefore be called a theory of the Electromagnetic Field, because it has to do with the space in the neighbourhood of the electric or magnetic bodies, and it may be called a Dynamical Theory, because it assumes that in that space there is matter in motion, by which the observed electromagnetic phenomena are produced (J. Clerk Maxwell, 1865, p. 460).

Maxwell formulated twenty equations (Figure 35) that were to become known as Maxwell's equations. They were later reduced, in 1884, by Oliver Heavide to the four equations known today (Figure 36). In 1873 Maxwell published his magnus opus: the *Treatise on Electricity and Magnetism*, and in the preface he wrote:

> *The fact that certain bodies, after being rubbed, appear to attract other bodies, was known to the ancients. In modern times, a great variety of other phenomena have been observed, and have been found to be related to these phenomena of attraction. They have*

Equations		Law
$e + \dfrac{df}{dx} + \dfrac{dg}{dy} + \dfrac{dh}{dz} = 0$	(1)	Gauss' Law
$\mu\alpha = \dfrac{dH}{dy} - \dfrac{dG}{dz}$ $\mu\beta = \dfrac{dF}{dz} - \dfrac{dH}{dx}$ $\mu\gamma = \dfrac{dG}{dx} - \dfrac{dF}{dy}$	(2)	Equivalent to Gauss' Law for magnetism
$P = \mu\left(\gamma\dfrac{dy}{dt} - \beta\dfrac{dz}{dt}\right) - \dfrac{dF}{dt} - \dfrac{d\Psi}{dz}$ $Q = \mu\left(\alpha\dfrac{dz}{dt} - \gamma\dfrac{dx}{dt}\right) - \dfrac{dG}{dt} - \dfrac{d\Psi}{dy}$ $R = \mu\left(\beta\dfrac{dx}{dt} - \alpha\dfrac{dy}{dt}\right) - \dfrac{dH}{dt} - \dfrac{d\Psi}{dz}$	(3)	Faraday's Law (with the Lorentz Force and Poisson's Law)
$\dfrac{d\gamma}{dy} - \dfrac{d\beta}{dz} = 4\pi p'$ $\quad p' = p + \dfrac{df}{dt}$ $\dfrac{d\alpha}{dz} - \dfrac{d\gamma}{dx} = 4\pi q'$ $\quad q' = q + \dfrac{dg}{dt}$ $\dfrac{d\beta}{dx} - \dfrac{d\alpha}{dy} = 4\pi r'$ $\quad r' = r + \dfrac{dh}{dt}$	(4)	Ampère-Maxwell Law
$P = -\zeta p \quad Q = -\zeta q \quad R = -\zeta r$		Ohm's Law
$P = kf \quad Q = kg \quad R = kh$		The electric elasticity equation ($E = D/\varepsilon$)
$\dfrac{de}{dt} + \dfrac{dp}{dx} + \dfrac{dq}{dy} + \dfrac{dr}{dz} = 0$		Continuity of charge

Figure 35: Maxwell's Equations in his original notation in *A Dynamical Theory of the Electromagnetic Field*

Source: http://www.ieeeghn.org/wiki/index.php/ STARS: Maxwell's_Equations

been classed under the name of Electric phenomena…Other bodies, particularly the loadstone, and pieces of iron and steel which have been subjected to certain processes, have also been long known to exhibit phenomena of action at a distance. These phenomena, with others related to them, were found to differ from the electric phenomena, and have been classed under the name of Magnetic phenomena…In the following Treatise I propose to describe the most important of these phenomena, to shew how they may be subjected to measurement, and to trace the mathematical connexions of the quantities measured. Having thus obtained the data for a mathematical theory of electromagnetism, and having shewn how this theory may be applied to the calculation of phenomena, I shall endeavour to place in as clear a light as I can the relations between the mathematical form of this theory and that of the fundamental science of Dynamics, in order that we may be in some degree prepared to determine the kind of dynamical phenomena among which we are to look for illustrations or explanations of the electromagnetic phenomena…It appears to me, therefore, that the study of electromagnetism in all its extent has now become of the first importance as a means of promoting the progress of science (J. Clerk Maxwell, 1873, p. Preface).

He did what he intended to do, understanding the relation between electricity and magnetism, and he explained the nature of electromagnetism in mathematical terms. His most prominent achievement was formulating a set of equations that united previously unrelated observations, experiments, and equations of electricity, magnetism, and optics into a consistent theory. His theory of classical electromagnetism demonstrates that electricity, magnetism, and light are all manifestations of the same phenomenon, namely the electromagnetic field.

1. $\nabla \cdot \mathbf{D} = \rho_v$
2. $\nabla \cdot \mathbf{B} = 0$
3. $\nabla \times \mathbf{E} = -\frac{\partial \mathbf{B}}{\partial t}$
4. $\nabla \times \mathbf{H} = \frac{\partial \mathbf{D}}{\partial t} + \mathbf{J}$

Figure 36: The four Maxwell's Equations
Source: www.maxwells-equations.com/

Maxwell's achievements concerning electromagnetism have been called the "second great unification in physics." At the time of his death of cancer at the age of forty-eight, in 1879, Maxwell's theory of electricity and magnetism was one of several. Its correctness was established only in 1887, when the German Heinrich Hertz discovered electromagnetic radiation at microwave frequencies, as Maxwell had predicted.

Heinrich Herz: electromagnetic waves

The phenomenon of the electromagnetic field also caught the attention of German physicist Heinrich Herz (1857–1894). He expanded Maxwell's electromagnetic theory of light by proving the existence of electromagnetic waves (for example light, a wave in the visual spectrum), then called "Herzian waves," as the result of electromagnetic radiation.

> *What Hertz did was very simple: he charged with electricity a "Leyden jar," which was a glass vessel with two plates of metal foil separated by air, an early version of the condenser or capacitor found in all electronic devices. This jar stored the electrical charge for a short time. Wires connected to the two metal plates were connected through a telegraph key to closely spaced electrodes so when the key was closed a spark appeared between the electrodes. On the other side of the room, a metal circle with both ends just about touching would "receive" or indicate a spark visually when one was "sent" from the Leyden jar, key, and spark gap.*

This needed to be improved upon, but it did confirm that Maxwell's theory had validity. Hertz proved that this unseen electrical force is transmitted through space and received without connecting wires, and that it had a wave-like nature that could be measured (Adams, 2012, p. 10).

This work done by Herz became the basis of (high frequency) radio transmission and the wireless telegraph. It was the Italian Guglielmo Marconi (1874–1937), child of an Italian/British couple, who pioneered long-distance radio transmission. His experiments proved that signals could be transmitted over "airwaves" using a transmitter and a receiver. And this was exactly what sparked Lee de Forest when he wrote his thesis on Herzian waves.

To conclude this brief overview of the scientific influences of electricity, we see that, over a considerable period of time, these *experimental* and *theoretical scientists* collectively unraveled the mystery of electricity (Figure 43). They made it clear what the nature of electricity was, how it could be created, and how it could be controlled. It was the sum of all the discoveries and inventions: such as the electric battery (by Alessandro Volta), the electric current (André-Marie Ampère), the electric charge (Charles-Augustin de Coulomb), electromagnetism created by electric currents (Hans Christian Oersted), the electric resistance (Georg Simon Ohm), the electromagnetic field creating electricity (Michael Faraday), the electromagnetic waves (Heinrich Herz), etc. Each of these discoveries was the result of other specific developments. And they were interrelated. Volta's discoveries were inspired by the work of Luigi Galvani (the frog's leg experiment), and he in turn influenced the work of Michael Faraday. Oersted's experiments stimulated many and influenced Ampère. Herz sparked the creativity of Lee de Forest…and so on. Sometimes the exchange of ideas was harmonious; other times differing ideas resulted in debates and controversies with accusations of plagiarism (Klotz, 1993).

Creation of electromagnetic power

Scientists had clarified the concepts of electricity and electromagnetism — more or less. But not everything was completely clear in the early nineteenth century. The next question was what to do with this new phenemenon. Could electricity be used to create motion: the electromotive power? The same question existed that was posed more than a century before as the question "Can steam create motion?" Indeed, just as steam proved to be usable to create a steam engine, the answer to this question, when applied to electricity, proved to be positive, and the result was that the electromotive engine was invented. However, the "invention" of the electric motor (and subsequently the electric dynamo) was not a single act by a single person at a certain moment in time (Gooding, 1985).

Many contributed to its development, both in Europe and the United States. Some of these advances are well known because they are well documented; other are lost in the fog of time. But there are some specific moments of importance that are distinguishable. The invention of the electromagnet is one of them.

William Sturgeon: the electromagnet

The Englishman William Sturgeon (1783–1850), was a self-educated man who was apprenticed to a shoemaker when he was ten years old—a master who starved and ill-used him. He ran away and joined the army in 1802.[40]

> *Seeing no hope of advancement in his trade, he enlisted in the Westmoreland militia, and two years later, being then twenty-one, he enlisted as a private in the royal artillery. His attention is said to have been directed to electrical phenomena by a terrific thunderstorm which occurred when he was stationed at Newfoundland. He determined to study natural science; but, finding himself unable to understand what had been written on the subject, he set himself, amid all the disadvantages of barrack life, to acquire the rudiments of an education. A sergeant lent him books, which he studied at night with the connivance of the officers; he is said to have ingratiated himself with the mess by his skill as a cobbler. In this way he worked at mathematics, and learnt sufficient Latin and Greek to grapple with scientific terminology.* (Grace's Guide)

---

[40] This information is based on biographies found at Grace's Guide (accessed January 2015): (http://www.gracesguide.co.uk/ William_Sturgeon), Wikisource (http://en.wikisource.org/wiki/Sturgeon,_ William_(DNB00), and Incredible People (http://incredible-people.com/biographies/ william-sturgeon/).

...he borrowed books to teach himself the basics of language, mathematics, and physics. He soon became popular with the cadets for his electric shock-inducing kites, and began to make scientific apparatus. Inspired by a bad thunderstorm, he began to investigate electrical discharges such as lightening, a study which he continued after leaving the army in 1820. For a time he resumed his old trade of boot maker, opening a shop. Here, during his leisure time, he taught himself turning and lithography, and devoted a good deal of attention to the construction of scientific apparatus. He supplemented his income by lecturing to schools and teaching officers' families...In 1824, he was appointed lecturer in science and philosophy at the East India Company's Royal Military College at Addiscombe in Surrey, England. He especially liked Oersted's experiment of 1820 because it linked electricity and magnetism for public entertainment and edification. In order to demonstrate electrical experiments, he needed equipment that was expensive and difficult to operate. While searching for affordable equipment, he invented the first practical electromagnet. (Incredble People)

Based on Faraday's concept of the electromagnetic field, William Sturgeon created the electromagnet in 1824–1825: a device existing of an iron core with a wire coiled around it. It could lift and hold pieces of iron due to its magnetic force (Figure 37).

Figure 37: Electromagnet developed by William Sturgeon (1824)
Source: Wikimedia commons

Sturgeon first applied his ideas of electromagnetism into a solenoid device. He wrapped several turns of wire around an iron core to produce magnetism when an electrical current was passed through the wire. He noticed that the electricity had set up a magnetic field that was concentrated in the iron core. He next varnished the iron to insulate it from the wound wires, and then hit on the idea of the horseshoe shape. He observed that each coil reinforced the next coil because they formed parallel wires with the current moving in the same direction. (Incredble People)

In 1825 Sturgeon presented to the Society of Arts the set of improved apparatus for electromagnetic experiments, including his first soft-iron electromagnet, for which he was awarded the silver medal of the society and a premium of thirty guineas. To him is undoubtedly due, says James Prescott Joule [q. v.], the credit of *being the original discoverer, he having constructed electromagnets in soft iron, both in the straight and horseshoe shape, as early as 1823...In 1832 he constructed an electromagnetic rotary engine, the first contrivance, according to Joule, by means of which any considerable mechanical force was developed by the electric current.*

In the England of that time, Sturgeon was considered to be an electrician, not a scientist. He was the founder of the Electrical Society of London, which was formed for the electricians of London in 1836 to serve as a forum to members and guests for reading and discussing papers on electrical experiments. Members of the Society gave public lectures at the Adelaide Gallery of Practical Science and often conducted research privately. The Electrical Society of London initially used the journal *Annals of Electricity* to report its activities and publish any science related to electricity. This was something different than the elite scientists who had organized themselves in the Royal Society of London. There, the famous "gentlemen of science," such as Robert Boyle and Michael Faraday, met in a more elite environment to discuss their findings and experiments.

Joseph Henry: the improved electromagnet

The American scientist Joseph Henry (1797–1878) was born in Albany, New York, to poor Scottish immigrants. As a small boy, he was sent to live with his grandmother. There he worked in a general store after school hours and, at the age of thirteen, was apprenticed to a watchmaker. As a young man, he became interested in the theater and was offered employment as a professional actor, but in 1819 several well-positioned Albany friends persuaded him instead to attend the Albany Academy, where free tuition was provided. His interest in science had already been aroused by a chance encounter with a popular scientific book, and by 1823 his education was so far advanced that he was assisting in the teaching of science courses. By 1826, after a stint as a district schoolteacher and as a private tutor, he was appointed professor of mathematics and natural philosophy at the Academy. In 1832 he became professor at Princeton University (Leitch & Leitch, 1978).

Henry was interested in terrestrial magnetism. In 1827 he visited New York City and attended a demonstration of the electromagnet devised by William Sturgeon. This stimulated him to make a better, more powerful, magnet (Figure 38). As

Figure 38: Joseph Henry's electromagnet (1831)

Source: Smithsonian Institution Archives
http://siarchives.si.edu

the electromagnet was switched on by applying an electrical current, the iron bar closing the magnetic circuit would stay firmly in place. That the electromagnetic field was creating a strong force was demonstrated by lifting weights of several hundred kilograms (Figure 39).

In experimenting with such magnets, Henry observed the large spark that was generated when the circuit was broken, and he deduced the property known as self-inductance, the inertial characteristic of an electric circuit. He published his findings (Henry, 1832), but these were published after Faraday had presented his own findings. Henry was slow in publishing his results, but he was unaware of Faraday's work. Today Faraday is recognized as the discoverer of mutual inductance (the basis of transformers), while Henry is credited with the discovery of self-inductance (Henry, 1839). His experiments with the electromagnet were focused on the creation of a motion (as demonstrated by his oscillating electromagnetic motor) or on the transmission of (or calling in action) power at a distance (as demonstrated by his doorbell).

Figure 39: Joseph Henry constructs large electromagnet for Yale College Professor Benjamin Silliman.

Source: Smithsonian Institution Archives

After Sturgeon and Henry, other scientists tried these experiments with electromagnets—for example the Dutchman Gerard Moll of the University at Utrecht, who in 1830 reported about a large electromagnet, weighing twenty-six pounds, that lifted thirty-eight kilograms (154 pounds) when excited by a battery.

Electrochemistry: the wet cell

For all of these experiments, the source of electrical energy was the "wet cell"—the battery in which an electrochemical process created an electrical current. All electrical research and its applications were based on this battery, and the early developments around the electromotive engine would not have been possible without the *electrochemical*

Figure 40: Cruickshank and the first flooded battery (1802)

Source: http://batteryuniversity.com/learn/article/when_was_the_battery_invented

battery. A device that underwent major changes after its basic phenomenon was discovered by Alessandro Volta in 1800. But before touching upon that development, we have to explain some basic 'chemical' characteristics of this device.

Electricity is based on the flow of electrons. Electrons are created in the chemical process within the wet cell. When the zinc anode and the copper cathode are connected by a wire, a current of these electrons starts to flow. Two things are happening:

1. The metallic zinc at the surface of the zinc electrode is dissolving into the electrolyte as electrically charged ions (Zn^{2+}), leaving two negatively charged particles called electrons (e^-) behind in the metal: $Zn \rightarrow Zn^{2+} + 2e^-$. This reaction is called *oxidation*.

2. While zinc is entering the electrolyte, two positively charged hydrogen ions (H^+) from the electrolyte combine with two electrons at the copper electrode's surface and form an uncharged hydrogen molecule (H^2): $2H^+ + 2e^- \rightarrow H_2$. This reaction is called *reduction*.

The electrons used from the copper to form the molecules of hydrogen are made up by an external wire or circuit that connects it to the zinc: the electrical current. The hydrogen molecules formed on the surface of the copper by the reduction reaction ultimately evanish as hydrogen gas. The main characteristic of the battery is the electrical current travelling from the point of high potential (the plus pole) to the point of low potential (the minus pole). For a zinc/electrolyte/copper cell, the difference in potential is 0.76 volts.

That is the technical explanation of the working of the galvanic battery, but there were other important characteristics such as its costs, heavy weight, and considerable volume. Volta's battery consisted of brine-soaked pieces of cloth sandwiched between zinc and copper discs, piled in a stack. This design resulted in electrolyte leakage as the weight of the discs squeezed the electrolyte out of the cloth. William Cruickshank solved this problem in 1802 by laying the battery on its side in a rectangular box (Figure 40). The inside of this box was lined with shellac for insulation, and pairs of welded-together zinc and copper plates were evenly spaced in the box. The spaces between the plates (the troughs) were filled with dilute sulfuric acid. So long as the box was not knocked about, there was no risk of electrolyte spillage.

Enough for the basics of the battery, now to its development over time. After Volta's discovery of the voltaic effect in 1800, a lot of scientists became interested—people such as Humphry Davy (1778–1829) and Andrew Cross (1784–1855). It was Davy who, in 1813, constructed a two-thousand-plate paired battery in the basement of Britain's Royal Society, covering eighty-three square meters (!). He surely was not the only one, as more scientists were interested in the new phenomenon of the voltaic cell:

The Englishman John Frederich Daniell (1790–1845), professor in chemistry, developed the Daniell cell in 1836 (Figure 41). This battery was a copper pot filled with a copper-sulfate solution. In this solution an earthenware container, filled with sulfuric acid and a zinc electrode, was placed.

Figure 41: Battery of six Daniell cells (1836)

Source: Wikimedia commons

Variations on this cell were the Bird cell, developed in 1837 by Golding Bird (1814–1845), the Bunsen cell developed by Robert Bunsen (1811–1899) in 1841, the Callaud cell developed by the Frenchman Callaud in the 1860s, and the Poggendorff cell developed by the German Johann Poggendorff (1796–1877). The Grove cell, a fuel cell, was invented by William Robert Grove in 1839. The Weston cell—which produces a highly stable voltage suitable as a laboratory standard for calibration of voltmeters—was invented by Edward Weston in 1893 (US patent №. 494.827).

The rechargeable battery was developed by the German Wilhelm Josef Sinsteden (1803–1891) and the Frenchman Gaston Planté (1834–1889) in 1859 (Kurzweil, 2010). They applied plates of lead sulphite ($PbSO_4$) and sulfuric acid (H_2SO_4) to create a chemical, reversible reaction: in the discharged state, both the positive and negative plates become lead sulfate ($PbSO_4$), and the electrolyte loses much of its dissolved sulfuric acid and becomes primarily water. Here is the reaction written in chemical terms: $Pb(s) + PbO_2(s) + 2H_2SO_4(aq) \rightarrow 2PbSO_4(s) + 2H_2O(l)$.

An avalanche of rechargeable batteries followed Planté's battery—for example the Leclanché battery (Figure 42). In 1866 Georges Leclanché patented a new system, which was immediately successful (US 55.441, June 5, 1866). In the space of two years, twenty thousand of his cells were being used in the telegraph system. Leclanche's original cell was assembled

Figure 42: Leclanché cell (1866)

Source: http://physicsmuseum.uq.edu.au/leclanche-cells

in a porous pot. The positive electrode consisted of crushed manganese dioxide with a little carbon mixed in. The negative pole was a zinc rod. The cathode was packed into the pot, and a carbon rod was inserted to act as a currency collector. The anode or zinc rod and the pot were then immersed in an ammonium chloride solution. The liquid acted as the electrolyte, readily seeping through the porous cup and making contact with the cathode material. Leclanché's "wet" cell (as it was popularly referred to) became the forerunner to the world's first widely used battery, the zinc carbon cell.

It was the electrochemical voltaic battery that would be the primary source of electrical energy for decades. It heavily influenced the development of electric devices. Not only in the nineteenth century, but even in our present time, the electrical battery is playing an important role. AS one can simply check by counting the numerous household artifacts that are battery-powered.

The power of lightning understood

As we have described before, science ended up with considerable knowledge about the fundamentals of electricity. Scientists and engineers managed to use this knowledge and apply electricity to daily life. Electricity that was used to create *linear power* (the electromagnet) and *rotative power* (the electromotive motor). These fundamental developments had an even larger impact than the steam engine (also with both the linear and rotative power created by the steam engine). Then it was steam that was the medium that transformed the "power of fire" into linear and rotative mechanical power. The basic component then was heated water (which created steam). Now electricity proved to be the next addition to steam—a new medium for the transformation of the "power of lightning" into mechanical power. The basic component was "heated" electrons (which created an electric current).

A legion of curious and inventive scientists worked hard to contribute to these developments—from the early electrophysicists who were discovering the underlying principles, to engineering scientists who were creating electric devices, to the theoretical scientists who were explaining the phenomenon (Figure 43). Some of them still quite know to the general public because their name was used to describe the phenomenon at hand: like André-Marie Ampere gave his name to the unit of electric current (amperes), Alessandro Volta gave his name to the electrical potential difference (volts). And George Ohm gave his name to the relation between volt and ampere: the resistance expressed in 'ohms'.

Consider the impact of the steam engine in the nineteenth century,[41] and it is not hard to understand what the impact of electricity was going to be. Similar to the static and mobile applications of the steam engine, the applications of electricity would be earth shaking. As we will explore in the next chapters.

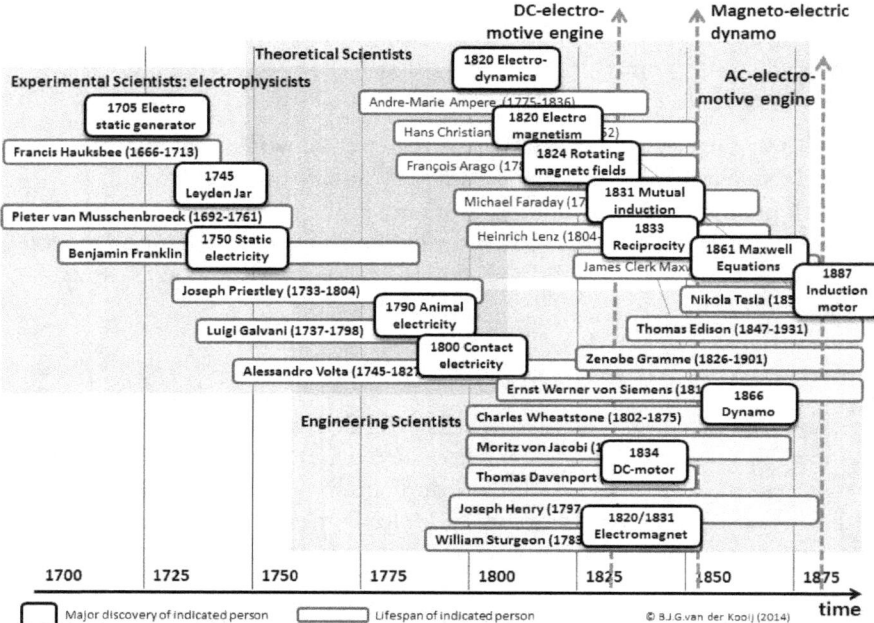

Figure 43: Scientists discovering, engineering and explaining electricity
Source: Figure created by author

---

[41] See: B. J. G. van der Kooij, *The Invention of the Steam Engine*. (2015).

The invention of the electric DC motor

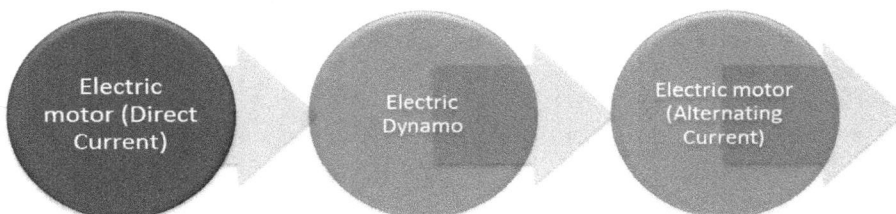

The experimenting scientists knew the basic mechanisms of electricity and had created their models in the early nineteenth century, but the application of electricity outside the scientific community into working products—such as the electric motor—took a while. Oersted, Faraday, Ampère, and others had shown the mechanism of electromotive power; it was a range of others' discoveries that brought the electric motor into the daily world. These important applications included the invention of the incandescent electric lamp and the electromechanical relay[42]. Each of these examples created an enormous impact in people's private and business lives.

The direct current electric motor

In the early days of development, the basic principles and ideas had to be transformed into working artifacts—even without the theoretical understanding available. It was a challenge for the many experimenters who were fascinated by the new phenomenon of electricity. Many of their efforts in the first half of the nineteenth century have disappeared in the fog of history, but some of their ideas got a foothold—such as the development trajectory that started with the linear movement created by the force of the electromagnet.

---

[42] See: B.J.G. van der Kooij: *The Invention of the Electric Light; The invention of the Communication Engines.* (2015)

The electromagnetic reciprocal engine

Many scientist/inventors were stimulated by the possibilities of the electromechanical force and tried to develop electromotive engines. At first the linear, reciprocating electric "common beam" motors were developed, using two powerful electromagnets—for example Henry's magnetic rocker (Figure 44), a philosophical toy created in 1831 by Joseph Henry. Although electromagnets now replaced the steam cylinders, the results had a striking resemblance to the first steam engines (Figure 45)!

Figure 44: Joseph Henry's oscillating beam electrical motor (1831)

Source: Smithsonian Institution Archives

Later this technology proved to be a dead end, but the equivalent development trajectory of the rotatory steam engine—the rotatory version of an electric engine using electromagnets—sparked the imagination:

Figure 45: Jean Bourboze's electrical engine imitating a steam engine (1865)

Source: www.earlyelectricmotors.com/

> *Public interest in the development of the electric motor received a great boost when, in 1835, M. H. Jacobi published a paper in which he argued that it should be possible to get enormous power from an ultra-high speed rotary electric motor and battery. The paper was translated into English and other languages with the result that a veritable euphoria swept Europe and the United States. The world was on the verge of very cheap power; the dirty, clumsy steam engine was to be replaced by the clean, compact, smoothly running electric motor. Steam locomotives were to be replaced by electric locomotives (battery driven) and, it was reported, one actually ran, briefly, on a main line in 1841…After the brief "electrical euphoria" interest in the electric motor lapsed until the end of the century. In the meantime the increasingly versatile, increasingly powerful steam engine met most needs for power so that it became the widely accepted symbol of the age* (Cardwell, 1992, p. 482).

The Page electrical motors (1840–1854)

One of the many inventors interested in the field of electromagnetism was Charles Grafton Page (1812–1868). After his graduation from Harvard Medical School in 1836, Page set up a medical practice in Virginia. In 1841 he became a patent examiner in the United States Patent Office (1842–1852 and 1861–1868) and professor of chemistry in the medical department at Columbian College (1844–1849)—now George Washington University—in Washington. His scientific work started with using electricity in medical applications: the early form of electrotherapy. During the 1840s he developed his Axial Machine: an electric motor with a reciprocating motion that was converted in a rotatory motion. Its motion very much resembled that of a steam engine's piston. Later, on January 31, 1854, he received US patent № 10.480 for his "Improvement in electro-magnetic engines" (Figure 46). The model of the machine to be patented shows a striking resemblance to a compound steam engine with a reciprocating piston engine.

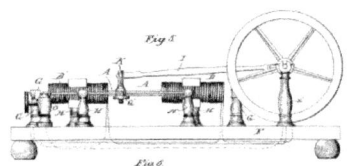

Figure 46: Patent 10.480 and model of Page's Axial machine (1854)
Source: USPTO

After demonstrating his machine's use in powering saws and pumps, he obtained funds from the US Senate to produce an electromagnetic locomotive. About 1850, Page received $20,000 in congressional support to build two special electromagnetic engines for a locomotive (Figure 47).

Figure 47: Drawing of Page's locomotive
Source: (Post, 1972, p. 141)

Professor Page made a trial trip with his electromagnetic locomotive on Tuesday April 20, 1851, starting from Washington, along the tracks of the Washington and Baltimore Railroad. His locomotive was of sixteen horsepower, employing one hundred cells of Grove nitric acid battery, each having platinum plates eleven inches square. The progress of the locomotive was at first so slow that a boy was enabled to keep pace with it for several hundred feet. But the pace soon increased, and Bladensburg, a distance of about five miles and a quarter, was reached, it is said, in thirty-nine minutes. When within two miles of that place, the locomotive began to run at a rate of nineteen miles an hour, or seven miles faster than the greatest speed theretofore attained. This velocity was continued for a mile, when one of the cells cracked entirely open, and, as a consequence, the propelling power was partially weakened. Two of the other cells subsequently met with a similar disaster. It was found that the least jolt, such as caused by the end of a rail a little above the level, threw the batteries out of working order, and the result was a halt. This defect could not be overcome, and Professor Page reluctantly abandoned his experiments in this special direction (Michalowicz, 1948, p. 1039).

The money Page received from Congress, something like an early scientific grant for science, had a negative side effect though. It caused Page to become an outcast among scientists in the United States because taking government money for research was considered unethical in 1850 (Currier, 1857, p. 2).

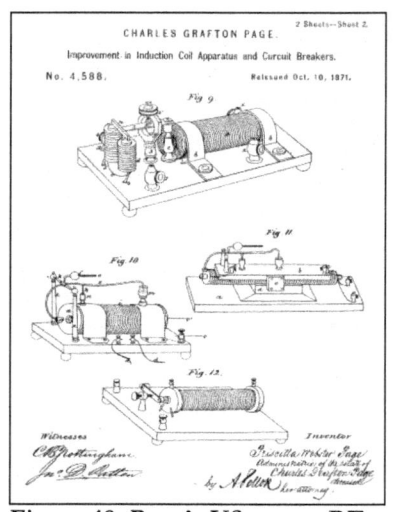

Figure 48: Page's US patent RE 4.588, October 10, 1871
Source: USPTO

In 1864 a heated debate started about who invented the electric motor. Emperor Napoleon III had awarded the prestigious Volta Award to Heinrich D. Ruhmkorff for *l'invention de la bobine d'induction*. Page claimed he had already invented the coil thirteen years earlier. This disagreement resulted in the involvement of the House Committee on Patents from the American Congress, as Page wanted a special act authorizing him to obtain a patent for his induction apparatus. In this case there was more than just a simple priority case; American pride was at stake—or as it was stated by the chairman of the Committee on Patents in the congressional debate in February 1868:

> *The purpose of the bill was "to protect the rights of an American inventor against the claims to originality, unjustly and mistakenly awarded to a foreigner by the high authority of a foreign power"* (Post, 1976, p. 1283).

Page succeeded. The Senate and the House of Representatives passed the special act, and it was signed by President Andrew Johnson. Page got US patent №. 76.654 granted on April 14, 1868, for his "Improvement in induction-coil apparatus and circuit-breakers" (resissued as RE 4.588 on October 10, 1871). His priority was established. Later, lawyers employed by Western Union—then owning half of the patent, the other half being owned by Page's widow Priscilla Webster Page—even started commencing infringement cases. However, Page was not involved anymore, as he had died on May 5, 1868, virtually penniless.

> On March 26, Bailey delivered Page's petition for a thirteen-claim patent to the Patent Office. On the thirtieth the application was approved, and the patent ordered to issue by the commissioner. Bearing number 76,654, it took effect on April 14. Page died just three weeks after, on May 5 (Post, 1976, p. 1284).

Other reciprocating electrical direct current (DC) motors

In Europe it was inventors such as the Italians Salvatore dal Negro (1768–1839), Luigi Magrini (1802–1868), and Guiseppe Botto (1791–1865) who created (reciprocating) electric engines. In 1830 Botto described in a note a prototype electric motor on which he was working. Around 1836 he published a description of it in a memorandum to the Academy of Turin titled *Machine Loco-motive mise en mouvement par l'électro-magnétisme*.

Figure 49: Botto's electrical motor (1834)
Source: Museo Gallileo, Florence

> Salvatore dal Negro of the university at Padua reported in 1834 on an invention that he had worked out in 1831 of a permanent magnet pendulum kept in oscillation by an electromagnet that changed its polarity by a commutator switch. He added a linkage device so that he could raise a weight with it and found it lifted sixty grams, five centimeters in one second. A similar pendulum-instrument was made in 1834 by J. D. Botto in Turin (King, 1962, p. 261).

J. J. Greenhough patented a reciprocal engine (GB Patent №. 13.613 filed on May 3, 1851). It took some decades before another patent was issued for this type of motor; on August 16, 1870, Landy Tunstall Lindsey was granted US patent №. 106.493. However, the "reciprocating electromotive" technology can be characterized as a dead-end technology as far as it concerns electrical motors.

Figure 50: Magrini's electrical motor (1840)
Source: Museo Gallileo, Florence

The electromagnetic rotatory engine

The principle of the electromagnet made the electric motor feasible (it was originally called the "electromagnetic engine").[43] The process of the electric motor's development was of an evolutionary nature as many engineers/inventors contributed to it.

> From the late 1830s through the 1850s the induction coil grew into a remarkable device with numerous improvements in design and construction. Several notable individuals of science played a prominent role in its development during this period of time such as: William Sturgeon (1783–1850), George H. Bachhoffner (1810–1879), Alexander Kemp, Christian Neef (1782–1849), James W. McGauley (1806–1867), Dr. Golding Bird (1814–1854), and Dr. Guillaume Duchenne. Of course others also made minor changes in the development of the induction coil, and these were mostly instrument makers such as: Edward Palmer, Watkins & Hill, and E. M. Clarke. Englishmen dominated the development of the induction coil (Currier, 1857, p. 4).

There were also those who tried to develop the reciprocal machines. William Ritchie (1830) created a device where the electromagnetic force was applied to create rotatory motion: the Revolving Electrical Motor. Charles Page made his Revolving Magnet (1840) and Daniel Davis his Thermo-electric Revolving Wire Frames (1842). All these devices were used more to demonstrate the principle of rotatory motion due to electromagnetism, than they were electric motors usable in a practical way.

Figure 51: Ritchie's motor (1840s)
Source: www.earlyelectricmotors.com/

The Davenport rotating electrical DC machines (1834)

Thomas Davenport (1802–1851), a blacksmith with little formal education, became fascinated in 1833 when he saw a demonstration of the lifting power of an electromagnet invented by Joseph Henry. He bought the magnet and started experimenting. He noticed that the on-off switching of the magnet could result in a linear motion. After much experimenting he

---

[43] The following terminology is used in relation in relation to classic electric systems: **Magneto-electric:** Motors with "permanent magnet" field systems; **Dynamic electricity**: Electricity generated by rotary motion; **Dynamo electric** (dynamos): Motors with "electromagnetic" or wound field systems, conversion of mechanical energy into electrical energy; **Electrodynamic:** Electric motor, conversion of electrical energy into mechanical energy. **Alternator:** Alternating-current generator

converted this linear motion of the electromagnet to a rotatory motion in 1834 (Figure 52).

Figure 52: Davenport's electric motor (1834)
Source: Edward W. Byrn, A. M.: *The Progress of Invention in the Nineteenth Century* (1900)

In July 1834, I succeeded in moving a wheel about seven inches in diameter at the rate of about thirty revolutions a minute. It had four electromagnets, two of which were on the wheel, and two were stationary and placed near the periphery of the revolving wheel. The north poles of the revolving magnets attracted the south poles of the stationary ones with sufficient force to move the wheel upon which the magnets revolved, until the poles of both the stationary and revolving magnets became parallel with each other. At this point, the conducting wires from the battery changed their position by the motion of the shaft; the polarity of the stationary magnets was reversed; and, being now north poles, repelled the poles of the revolving magnets that they had before attracted, thus producing a constant revolution of the wheel (Michalowicz, 1948, p. 1036).

Davenport took his device to Middleburg College and showed it to the professors Turner and Fowler, who encouraged him to continue working on the engine and apply for a patent. He prepared a patent model (Figure 52), filed his claim on January 24, 1837, and received on February 25—only thirty days later—US patent №. 132 [44] for "Improvement in propelling machinery by magnetism and electro-magnetism" (Figure 53). The claims of his patent were extremely broad:[45] "Applying magnetic and electromagnetic

---

[44] Davenport was issued US Patent No. 132, because the 9,957 patents issued between July 31, 1790, and July 13, 1836, weren't numbered.
[45] This is similar to the patent on the "acoustic telegraph" (telephone) by Alexander Graham Bell (US patent 174.465 on March 7, 1876: Improvement in telegraphy) that claimed the broad area of "1….vibrations with undulatory currents of electricity…2….the combination of a permanent magnet with a closed circuit…3. The method of producing undulations in a continuous voltaic current by the vibration or motion of bodies…4. The method of producing undulations in a continuous voltaic current by gradually increasing or diminishing the resistance…5. The method of, and the apparatus for transmitting vocal or others sounds telegraphically…" etc. Another example would be Edison's patent for the incandescent lamp (US patent 223.898—granted on January 27, 1880) that claimed "any electric lamp for giving light by incandescence, consisting of a filament of carbon…the combination of carbon filaments with a receiver entirely made of glass and conductors…the method of securing the platius contact-wires to the carbon filament…" Both Bell and Edison were becoming rich men as a result of their discoveries, the patent, and its claims.

power as a moving principle for machinery in the manner above described, or in any other substantially the same in principle." But his patent brought him no financial reward. He died penniless in 1851, his patent just having expired (Michalowicz, 1948, p. 1039).

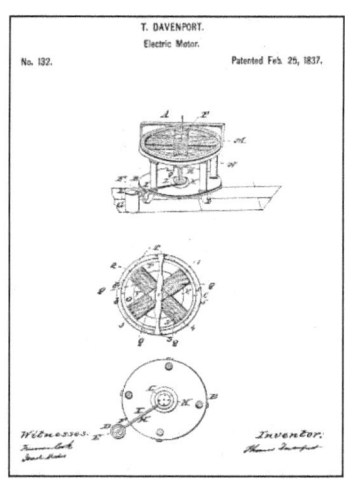

Figure 53: Davenport's patent No. 132 for an electric motor (1837)
Source: USPTO

> In an exhibition in London in August 1838, one of Davenport's motors drove a small electric train of several carriages with a total weight of seventy to eighty pounds at a speed of three miles per hour. Davenport tried to use his rotating motor to drive a Napier printing press that printed his paper, The Electro-Magnet, but the press required an engine from one to two horsepower, and he did not succeed in building such a motor until 1840. Success came to Davenport with his development of a reciprocating engine based on a "sucking coil" that he had begun working on in 1838. Davenport built over one hundred motors in his lifetime, but lack of financial backing and his inability to obtain an inexpensive source of power defeated him (King, 1962, p. 265).

Von Jacobi's electrical DC motor (1835)

It was the German Moritz von Jacobi (1801–1874) who developed a rotating electric motor. It was the result of his work in St. Petersburg on rotation by electromagnetic methods.

> Machines had fascinated Moritz Jacobi since his student days. The son of a wealthy Jewish merchant family in Potsdam, he had initially followed his parents' wish and studied civil engineering as part of Kameralistik in Berlin and at Göttingen, finishing a degree in architecture. Between 1825 and 1832 he worked as a Prussian civil servant, translated

Figure 54: Von Jacobi electric motor (1834)
Source: Wikimedia Commons

civil engineering books (Baukunstbücher), and finally became a civil engineer (Baumeister) in Königsberg where his younger brother had been professor of mathematics since 1827. But he did not stay for long. Already in 1835 Moritz Jacobi was elected professor of civil engineering (Zivilbaukunst) at the Russian University of Dorpat. There he designed and constructed major buildings in the city and the university. In the same year his Mémoire sur l'Application de l'Électro-Magnétisme au Mouvement des Machines (Jacobi 1835) was published in Potsdam, plans for which he had made as a student in Göttingen (Otto Sibum, 2003, p. 101).

He also presented the paper before the Academy of Sciences in Paris. In this paper, *Mémoire sur l'Application de l'Electro-Magnétisme au Mouvement des Machines*, he describes his machine. Independently from Davenport he discovered electromechanical motion (Figure 54).

With financial support from the Russian government, he started his trial project—a boat powered with his electrical machine—to prove that his model could be converted into a working machine. After two trial boats, he managed to create a third boat that was first tested on September 13, 1838, and in 1839 he gave a second, improved performance (Figure 55).

Figure 55: Engraving of an imaginary scene of Jacobi's boat experiment from 1838 to 1839 on the Newa in St. Petersburg
Source: (Otto Sibum, 2003, p. 106)

In the year 1839, the Emperor Nicholas of Russia granted a sum of $12,000 [46] *to De Jacobi to enable him to prove that his electric motor had practical application. De Jacobi had a boat constructed, twenty-eight feet long and seven feet wide, which was propelled by means of paddles connected to an electric motor of his design. The De Jacobi boat, the first practical application of an electric motor, carried about fourteen passengers and was powered by 320 Daniell cells, which is equivalent to about one hundred of our present-day six-volt storage batteries. In its trip up the Neva River, it never achieved a speed greater than three miles an hour and*

---

[46] This project amount would be the equivalent of more than $5 million in 2010, calculated on the basis of labor cost. Source: Measuring Worth at http://www.measuringworth.com/uscompare/relativevalue.php.

consequently offered very little competition to man-propelled craft. The weight of the many batteries made the accomplishment impractical, and it shortly was declared a failure. The cause of the failure was obvious, but the value of the electric motor as a new form of marine power was not forgotten (Michalowicz, 1948, p. 1038).

So Moritz Jacobi, called the "electromagnetic watt" by his brother, proved that he could turn his model into a working machine (Otto Sibum, 2003, p. 109).

Other developments of the rotatory DC electric motor

There were numerous other contributors to the creation of an electromagnetic rotative motor. Among those were: James Joule (English, 1838), William Taylor (English, 1838), Uriah Clarke (American, 1840), Thomas Wright (English, 1841, GB patent 9.204), Wheatstone (English, 1841), de Harlem (1841), P. Elias (American, 1842), G. Froment (French, 1844), Moses G. Farmer (American, 1846), G. Q. Colton (American, 1847), Sören Hjorth (Swedish, 1849, 1851), Thomas Hall (American, 1850), T. C. Avery (1851), Du Moncel (French, 1851), Marié Davy (French, 1855), Pacinotti (Italian, 1861), and others (Doppelbauer, 2012). Some are more known than others, though.

> The American Moses D. Farmer (1820–1893) of Dover, New Hampshire, devised an electric motor in 1846 that powered an electric train of two cars on an eighteen-inch-gauge track in its first public exhibition in July 1847. Farmer had other exhibitions in New England later that year; but his exhibitions were not financially successful, so he turned to the field of telegraphy (King, 1962).

> The Englishman Charles Wheatstone (1802–1875), professor at King's College, London, was one of the persons intrigued by von Jacobi's description of how his electromotive engine had powered a boat, and he started working on an electromagnetic machine. He developed three different machines and received, in 1841, a British patent №. 9.022 for his eccentric machine (Bowers, 1972) (Figure 56). This was one of the numerous patents that were issued for electromagnetic motors.

Figure 56: Wheatstone's eccentric ring type electromagnetic engine (1841)

Source: Science Museum/Science & Society

Paul-Gustav Froment (1815–1865) was a French engineer who devised an electric rotative motor in 1844 that was one of the first that was used for industrial purposes (Figure 57). He also created a reciprocal engine: the *moteur a piston electomechanique*.

Others in France were active too—for example A. J. L. H. Tourteau, Compte de Septeuil, who patented his machine in Britain (GB 840, November 24, 1852) (Dredge, 1882). The rotative DC electric motor had gained the most attention, and the reciprocating engines disappeared completely.

Figure 57: Electromagnetic engine, Gustav Froment (1845)
Source: www.earlyelectricmotors.com/

> *By this time two basic forms of the electric motor had been developed. One of the basic forms was a reciprocating engine, where an armature was pulled into a solenoid, as in Page's motor, or an armature hinged at one end was pulled down by an electromagnet, as in Clarke's motor. Linkages changed the linear motion to a rotary one. The other basic form was a paddle wheel, where an armature was kept in constant motion by a commutator switching on a field to tease the armature ahead at the right time. The engines of Ritchie, Jacobi, Davenport, Davidson, and Froment were of this second form. After midcentury there was a further proliferation of electric motors, but no new basic types were introduced until the advent of AC power* (King, 1962, pp. 268-269).

As can be concluded from the preceding comment, the rotative DC electric motor was slow in its early development—partly because its electric-power supply came from a battery, a technology still in its infancy in those days. And the rotative DC electric motor had a formidable competitor in the steam engine that became widely accepted as a source of reciprocating and rotatory power.[47]

> *The greatest difficulty in the use of electricity lay in the relatively high cost of production of electrical power in comparison with that of steam. Instead of consuming coal in a chemical reaction that produced heat and the expansion of water, one dissolved a metal in an acid in a chemical reaction that produced an electrical current. Metals and acids were much more expensive than coal and water…Another very important deterrent to the use of electrical power was the*

---

[47] See: B.J.G. van der Kooij: *The Invention of the Steam Engine.* (2015)

problem of distributing electrical current. Although by midcentury one could signal over long distances, power could be transmitted efficiently only within an area the size of a large room. Until some better means of distributing electricity was found, inventors had to use very bulky containers full of corrosive liquids directly at the place where the power was consumed (King, 1962, pp. 269, 270).

As the batteries powering the DC motors were impractical, bulky, and needed to be refueled by changing the metal plates and the acid regularly, the DC electric motor could not find commercial markets for its application—except a few limited opportunities such as the electric fan—and it more or less disappeared from the scene for some decades. "The revival of electric power had to wait for such developments as the invention of the self-excited dynamo, the application of electric light to lighthouses and then to streets and squares, and finally the masterly invention of the incandescent glow lamp by Edison and Swan" (Cardwell, 1976, p. 685).

The mere fact that electrical energy from the least expensive chemical battery using up zinc and acids costs twenty times as much as that from a dynamo—driven by steam engine—is in itself enough to explain why so many of the electric arts lingered in embryo after their fundamental principles had been discovered. Here is seen also further proof of the great truth that one invention often waits another (McPartland, 2006, p. 84).

The invention of the DC motor

As the preceding overview shows, many inventors contributed to the development of this first version of the "electromagnetic engine." That being said, one could ask oneself, "Who is the *inventor* of the DC engine?" It's hard to say, but is there a single distinguishable *invention* that marks a breakthrough and can be considered as *the* invention of the electric motor?[48] The answer lies in an invention completed in the same time frame. It was Thomas Davenport who was the first American to create a usable electric motor around 1834, and he got the first US patent for such a device in 1837. Von Jacobi developed his electric motor at the same time, independently, on the other side of the world in Russia. He later published his findings in 1835, and his idea was not patented. However, he demonstrated that his motor worked with his paddle-wheel boat demonstration in 1838or 1839 on the Newa in St. Petersburg.

---

[48] Franklin Leonard Pope, "The Inventors of the Electric Motor," *The Electrical Engineer* 11 (7 January 1891).

> *Davenport has the honor of being the first of thousands of engineers who received a patent for an electric motor. But he is neither their inventor nor did his designs have any significant influence on the further development of electric motors…Jacobi expressly claimed in the memorandum of 1835 that he was not the sole inventor of the electromagnetic motor. He indicates the priority of the inventions of Botto and Dal Negro. However, Jacobi is undoubtedly the first to create a usable rotating electric motor* (Doppelbauer, 2012).

So Davenport and Jacobi arrived at similar conclusions during the same time period, but they were unaware of each other's efforts.

> *On December 1, 1834, a Russian scientist, Moritz Hermann De Jacobi, presented a paper before the Academy of Sciences in Paris in which he stated that he had obtained rotation by electromagnetic methods in May 1834. Since Davenport wrote in his memoirs that he first succeeded in producing motion electrically in July 1834, the question of actual priority in the point of time, is a close one. If the results obtained by De Jacobi in May 1834 were affected by the same apparatus that he described in his paper before the Academy in December 1834, then the priority must be conceded to him, but there is no proof available to show that such is the case. However, there is no doubt that the discoveries made by the two men were wholly independent* (Michalowicz, 1948, p. 1038).

Both inventions proved that the principle idea of an electromotive engine could be realized. Neither of their efforts to convert the concept into a working product was that successful though. Other developments had taken place in England, Italy, France, and Germany, but not all those designs or models of electrical engines resulted in usable machines.

> *The efforts of these early pioneers, although seemingly failures, laid the necessary foundation for the development of the modern electric motor. Their accomplishments proved disappointing not because the soundness of the electric motor as a means of converting energy was in doubt, but rather because the device was born too soon…It was not until 1886 that the electric motor found its place in the American home when the Curtis, Crocker, Wheeler Company began manufacturing battery-operated motors for sewing machines. It was not until 1887 that the city of New York decided to lay the first tracks for its famous "elevated." It was not until 1897 that the first electric automobile made its appearance. It was not until 1910 that the first electrically driven washing machine was introduced to the American housewife. And it was not until 1915 that the USS Mexico, the first electrically propelled battleship, was launched* (Michalowicz, 1948, p. 1039).

From a *technical point* of view, one can conclude that Davenport and Von Jacobi proved that the DC motor was feasible. The steady improvement of electromagnets, made by numerous people such as Sturgeon and Henry, et al., formed the basis for the rotatory engines developed by so many experimenters. This early work more or less culminated in the invention of

Table 1: Overview of some patents for the electric DC motor (1837–1856)

Patent №	Year*	Patentee	Description
US 132	February 25, 1837	Th. Davenport	Improvement in propelling machinery by magnetism and electromagnetism
US 809	June 27, 1838	N. Wakly	Electromagnetic machine
GB 7.729	July 11, 1838	L. C. Callett	Propelling vessels, carriages / commutators
US 910	September 12, 1838	S. Stimson	Improved apparatus for the application of electromagnetism as a motive power: twelve pole
US 1.735	August 25, 1840	T. Cook	Improvement in electromagnetic engines: armature
GB 8.937	April 27, 1841	W. Petrie	Obtaining motive power / electric motor
GB 9.022	July 7, 1841	C. Wheatstone	Regulating and applying electric currents / electric motor
GB 9.053	August 21, 1841	F. De Moleyns	Production or development of electricity and its application to illuminating and motive purposes / electric motor
GB 12.295	October 26, 1848	S. Hjorth	Electromagnetism as motive power / electromagnetic motor
US 7.287	April 16, 1850	J. Lillie	Improvement in electromagnetic engines
US 7.889	January 7, 1851	J. Neff	Improvement in electromagnetic engines
US 7.950	February 25, 1851	Th. Avery	Improvement in electromagnetic engines: producing, regulating, and applying electric currents / electric motors
GB 13.613	May 8, 1851	J. J. Greenghough	Obtaining and applying motive power
US 10.480	January 31, 1854	C. G. Page	Improvement in electromagnetic engines
GB 148	January 19, 1855	P. A. Le Comte de Fontaine Moreau	Obtaining electromotive power / Magneto-electric motors
US 14.682	April 15, 1856	M. Vergnes	Improvement in electromagnetic engines: electro-galvanic machine for producing motion by galvanic electricity

*) Year: for GB patents—filing date, for US patents—granting date.

Source: USPTO. (Dredge, 1882); Appendix A: Abstracts of patents. Center for Research Libraries http://dds.crl.edu/loadStream.asp?iid=17444&f=8 (Accessed November 2014)

the DC motor. In this respect one can conclude that the work of both Davenport and von Jacobi certainly merit labels of being "the inventor" of the DC electric motor.

Their work, however, did not have a real *impact* that resulted in further market and business development. The early DC electric motors were a stagnant development because of their power supply, the cumbersome "wet battery." The result was that the DC motor would remain dormant for decenniums and would only reappear decades later in other dedicated applications (i.e., Sprague with his DC motor for rail carriages, Curtis, Crocker, Wheeler motors) when DC electricity would be available in abundance. Before the 1850s its development was stalled due to the limited supply of electric energy that the wet battery so awkwardly supplied.

Patent activity

The described activities, experiments, and developments have resulted in a range of patents indicating innovative activity. The totality of the DC motor was improved in parts—such as the improvements of the electromagnets, the commutators, and the armatures holding the coils. And they were related to the development of the electrtrodynamic generator. Due to the reciprocal nature between the motor and dynamo, many inventors occupied themselves with both fields. Some of the patents related to the DC motor are shown on the previous page in Table 1. They cover the electromagnets, commutators, and motors themselves. This none-too-complete overview shows the early patents related to the DC motor powered by batteries.

A cluster of innovations for the DC motor

As can be concluded from the preceding, the electrophysicists' discoveries created worldwide excitement. They resulted in numerous engineering scientists trying to apply the electromagnetic force and create movement. Originally, their inventions were focused on an oscilating movement that mimicked the steam engine with its piston-and-crank system; the early electric motors were reciprocal. The movement of a piece of iron in an electric coil was translated in a linear movement that became rotative with the crank. Although much effort was applied over the decades, the concept did not work, and this trajectory proved to be a dead end.

The development trajectory was different for those who chose to create rotative movement directly using the basic properties of electricty. By applying additional devices, such as an early kind of communtator, it proved to be possible to create a rotary movement. This resulted in the described efforts of both Davenport in the United States and Von Jacobi in Russia.

Whatever the case about priority, it is clear that both men showed that electromagnetism could be used to create a functional electromagnetic engine that could power an object (boat, train). Once the concept of the DC motor was established, it led to a rush in developmental activities all over the world. A range of DC motors was developed. As noted, this resulted in a lot of patent activity. But the inventors all had the same problem; they were based on the power supply of a wet battery. So their field of application was limited to low-power requirements. That would change when the electic dynamo was invented and would create an abundance of electrical energy (Figure 58).

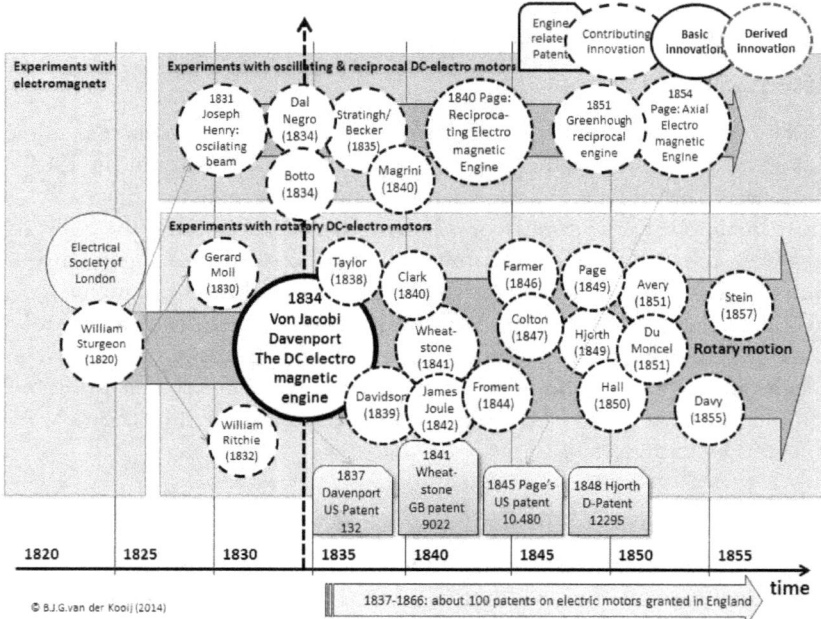

Figure 58: Cluster of Innovation around Davenport's and Von Jacobi's DC-electromotive engine
Source: Figure created by author

The invention of the electric dynamo

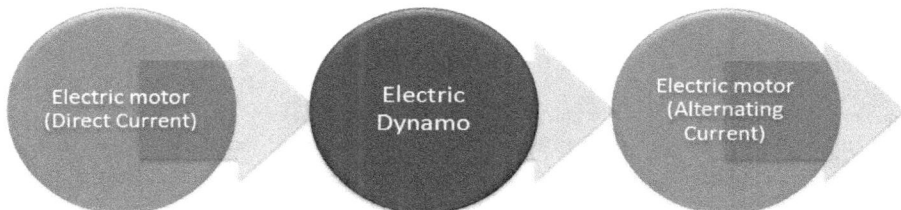

In the early years of electrical engineering, there was a distinction between the *magnetic-electro* machines (i.e., electric generators) and *electromagnetic* machines (i.e., electric motors). The main focus in the beginning of the nineteenth century was on electromotive power when using electricity to create motion. These developments were hampered by the lack of a useful source of electricity, as the electrochemical batteries were limited in their performance (in terms of capacity, weight, price, and maintenance). But there was a specific attribute of electromotive power that enabled its rise in popularity: reciprocity. The interesting phenomenon of the *electric motor* is that it has a "reciprocal" equivalent, the *electric dynamo*. The electric motor and the electric dynamo are, in fact, each other's opposite: The electric motor creates rotation energy from "electric power;" the dynamo creates "electric energy" from rotation power. Thus when one hooks up a steam engine with an electric dynamo, one has a manageable source of electrical energy.

It was this attribute that would change the future of electricity, as the electric dynamo[49]—together with the evolving infrastructure to distribute

---

[49] A little technical explanation could increase the reader's understanding of the basics of the dynamo: The dynamo uses rotating coils of wire and magnetic fields to *convert mechanical rotation into a pulsing electric current,* as the result of magnetic induction. A dynamo machine consists of a stationary structure, called the **stator**, which provides a constant magnetic field,

the electricity—would prove to be a reliable source of energy. It was the fundamental property of an electrical current to be able to span long distances that outperformed the capability of steam to deliver power over any distance. Steam engines were local by definition, and their power could only be distributed by a mechanical infrastructure of belts and line shafts. But electricity could be distributed over bigger areas by cables—certainly after the high-voltage networks of AC electricity were developed.

The wet battery had none of these attributes. It was by definition "local," its capacity was limited, and its operation impractical and expensive. The electrochemical battery seemed to create a barrier for the further penetration of electricity in practical applications—a barrier that would be demolished only by the arrival of the electric dynamo. The "dry" battery was the solution to the limitations of the "wet" battery.

Electricity generators: the dry battery

It was the Russian[50] Heinrich Friedrich Lenz (1804–1865) whose experiment with the direction of a current depending on the rotation of the magnet caused him to first realize the existence of the reciprocal property of electricity. Not only was it a property of "electricity creates motion," but there was also the reciprocal effect that "motion creates electricity." He described this effect in 1833 by formulating Lenz's law: i.e., the reversibility of the electric generator and the electric motor.

Lenz's explorations of electricity have to be seen as a step toward full understanding of electricity. In addition to the English, Danish, and Italian scientists mentioned earlier, many German and east European scholars contributed to the understanding of this new phenomenon of electricity.

> *The early years of the century saw a flurry of activity on the voltaic pile and various associated electrostatic and electrochemical phenomena by men such as Ritter, Erman, Jaeger, Pfaff, and, in the 1810s, Schweigger. With the discovery of electromagnetism in 1820 came contributions by Seebeck, Muncke,*

---

and a set of rotating windings called the **armature** (also **rotor**) which turn within that field. The motion of the wire within the magnetic field causes the field to push on the electrons in the metal, creating an electric current in the wire. On small machines the constant magnetic field may be provided by one or more **permanent magnets**; larger machines have the constant magnetic field provided by one or more **electromagnets**, which are usually called field coils. The field coils of the stator may be self-excited, using current generated by the dynamo itself, or separately excited by a separate, smaller, dynamo.

[50] Lenz was a Baltic German, and he was born in the Russian province of Livonia. The term "Estonia" during his entire lifetime referred to the Russian province of that name, which is the province north of Livonia, with the capital of Riga (now Tallinn). This means that he was an ethnic German but a Russian citizen.

Poggendorff, Pohl, and Schmidt, in addition to continued work by Erman, Pfaff, and Schweigger. These individuals set the tone for the study of electricity in Germany during the first quarter of the nineteenth century (Caneva, 1978, p. 64).

One has to realize this was in the early days of the scientific exploration of electromagnetic phenomena. Faraday had not long before published his findings on electromagnetic rotation (c. 1821) and electromagnetic induction (c. 1831). Lenz published his findings in 1834, in a paper titled: *Ueber die Bestimmung der Richtung durch elektodyanamische Vertheilung erregten galvanischen Ströme* (Lenz, 1834). He acknowledged that his work was a direct result of Faraday's publications.

> *Gleich bei Durchlesung der Abhandlung Faraday's schien es mir, als müssten sich sämtliche Versuche der elektrodynamischen Verteilung sehr einfach auf die Satze der elektrodynamischen Bewegungen zurückführen lassen, so dass, wenn man diese als bekannt voraussetzt, auch jene dadurch bestimmt sind, und da sich diese Ansicht bei mir durch vielfache Versuche bestätigt hat, so werde ich sie im Nachfolgenden auseinandersetzen, und theils an bekannten, theils an eigens dazu angestellten Versuchen prüfen*
> (Lenz, 1834, pp. 484-485).[51]

So Lenz studied electromagnetic phenomena, experimented with it, and concluded that an electromagnetic force always creates a counteracting force. This became Lenz's law that states that whenever a change in a magnetic field occurs, an electric field is generated to oppose the change. It relates to Newton's famous third law: for every action there is an equal and opposite reaction. Lenz's law provided a qualitative understanding of how a changing magnetic field creates an electric field. Turning a solenoid—an action that equals rotative motion—in a magnetic field creates an electric current, and applying an electric current in a magnetic field creates a (rotative) motion. Thus in a device like a dynamo, applying rotation power creates electricity (the electromagnetic force: EMF), but the rotation itself creates a counteracting electric current (back EMF).

The idea of using rotative motion to create electricity had far-reaching consequences. The resulting dynamo (called the "dry battery") was going to replace the electrochemical battery (the "wet battery"). By nature the wet

---

[51] Translation by author: "It seemed to me, the moment I read Faraday's treatise, as if all attempts by the electrodynamic distribution would very simply be traced back to electrodynamic movements, so that, if one assumes these as known, and because this view through many trials was confirmed to me, I will explain it in the following, partly on the known, partly by my own specific experiments."

battery had to be located close to its user; it was always a local source of electric energy. In contrast the dry battery could cover a larger area, as the supply of electric energy to the users would be realized by a distribution infrastructure: the electric grid. The availability of electric energy would allow light in every home and office, and it would enable the widespread use of the electric motor.

Principle of the dynamo: DC/AC

At first the electric dynamo became an important alternative to the battery, which had its operational limitations. In the same way that batteries were a direct-current (DC) device, the first dynamos were also DC devices. Basically the dynamo is a simple device: a coiled wire (the rotor) is turned between magnets, creating a magnetic field (the stator). The result is an (alternating) electric current (in short AC) as illustrated in Figure 59. All mechanical devices that produce electricity from a rotating magnetic field or from conductors rotating past the north/south poles of fixed magnetic fields produce *alternating current*. Using a commutator, the alternating current is converted to direct current (in short DC). The result is a DC current as illustrated in Figure 60. It is the "switching" commutator that makes the difference. However, it is a troublesome device in generator and motor design: it sparks, burns, and fails the system.

This basic difference in types of electric current would be the cause of a major battle between those who promoted the DC systems, and those who promoted the AC systems. When the DC systems came into existence, they

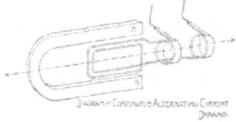

The coil is moved between the poles of a horseshoe-shaped magnet: N and S. The two separate metal contacts conduct the current from the two rings connected to the coil ends. The result of the rotation of the coil is a sinus-shaped voltage and current.

The coil C is moved between the poles of a horseshoe-shaped magnet: N and S. The metal contact A and B conduct the current from the coil ends R and T. The result of the rotation of the coil is a voltage shaped like a series of camelbacks.

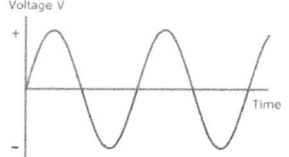

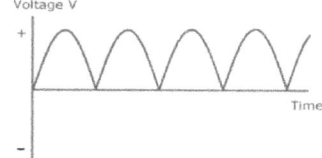

Figure 59: Principle of an AC dynamo

Figure 60: Principle of a DC dynamo

worked well in local systems, combined with storage batteries. However, due to the drop in voltage when increasing distance, they had a limited range. AC systems, using transformers to create high voltages, did not have that loss of power and could be used over longer distances, enabling larger distribution networks.

In the "Battle of the Currents" (see page 217) that resulted in the initial years of electricity distribution, the major adversaries were Thomas Edison and George Westinghouse—Westinghouse being supported by his inventor Nicola Tesla (see page 172).

Early magneto-electric dynamos

As stated earlier many scientists and engineers who worked on these magneto-electric machines were continuing in the footsteps of Michael Faraday's work on electromagnetic induction. In universities in many countries, scholars tried to experiment with the new toy: electricity.

Figure 61: Anyos Jedik's dynamo (1823)

Source: http://www.omikk.bme.hu/

One of these scholars was the Hungarian Benedictine Anyos Jedik (1800–1895). Around 1823 he formulated the concept of the dynamo (rotor, stator, and commutator) and created a dynamo using two electromagnets opposite each other to induce the magnetic field around the rotor (Figure 61).

Figure 62: Hippolyte Pixii's dynamo with the rotating magnets and the commutator (1832)

Source: Museo Gallileo, Florence

In France it was the Frenchman Hippolyte Pixii (1808–1835), an instrument maker in Paris, who created an early form of the dynamo in 1832, following the instructions of Ampère (Figure 62). He rotated the magnets and kept the coils stationary. He also applied a commutator, a rotary switch with contact bars that periodically switched the currents, thus creating DC electricity. Others working on the early dynamo were Saxton (who invented his 1833 machine), Uriah Clarke (with an 1835 machine, Figure 63), Woolrich (an 1841 machine), and Stoehrer (an 1844 machine). Also some reciprocating dynamos were made so that the magnet was moved in and out of a coil by cranking a handle.

The Italian professor in physics, Antonio Pacinotti (1841–1912), developed a dynamo in 1860 using a ring armature, later to be known as the "Ring of Pancinotti" (Figure 64). Instead of one coil, Pancinotti used multiple coils and adapted the commutator accordingly. The DC current the dynamo produced was a nearly nonfluctuating current. He published his discovery in *Il Nuovo Cimento*, June 1865, and exhibited the machine at the Paris International Exposition of Electricity of 1881. It was this Pancinotti ring that inspired the Belgian Engineer Zenobe Gramme (1826–1901) to create his dynamo in 1871 that used many armature windings, wound on a doughnut-shaped armature, and switched with a many-segmented commutator, to smooth the output waveform, producing nearly constant DC power.

Figure 63: Clarke's dynamo with the rotating magnets and the commutator (1835)

Source: Museo Gallileo, Florence
http://catalogue.museogalileo.it/

This is a sampling of the contributions from scientists all over Europe. Although language and geographic problems were quite substantial, it is amazing how news of discoveries travelled all over Europe. Oersted's (using Latin) and Faraday's (using English) publications travelled fast; other works (like Lenz's work in Russian) took more time to disperse in the scientific community.

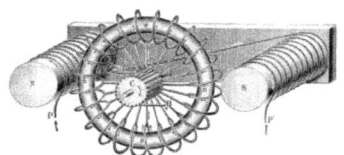

Figure 64: The ring of Pacinotti as used in his dynamo (1860)

Source: Wikimedi Commons

> *Les articles originaux de* LENZ *rédigés soit en russe, soit en allemand ont été publiés dans les comptes-rendus de l'académie de Saint-Pétersbourg, ou dans les annales Poggendorf. De ce fait, aux délais habituels entre la présentation du travail à l'académie et sa publication s'ajoutent parfois plusieurs années avant que le travail ne parvienne à Paris ou à Londres* (Khantine-Langlois, 2005, p. 707).[52]

---

[52] Translation by author: "The original Lenz articles, written either in Russian or German, were published in the proceedings of the Academy of St. Petersburg, or in the *Poggendorf Annals*. As such, the usual deadlines between the presentation of the work at the Academy

Dynamo: a developmental trajectory

The development of the generator over the years followed a trajectory that was realized—speaking in general terms—in the following steps:

Step to multicoil application: As explained earlier the magneto-electric machine is about moving a coil in a magnetic field—as shown in Figure 65. Using two coils results in a fluctuating current (top graph). By adding coils and poles, the result becomes a smoother curve, as shown below in the graph of the combined voltage created by the three poles (each 120 degrees out of phase from the other). Using more coils flattens out the fluctuations (bottom graph), resulting in a more steady power output.

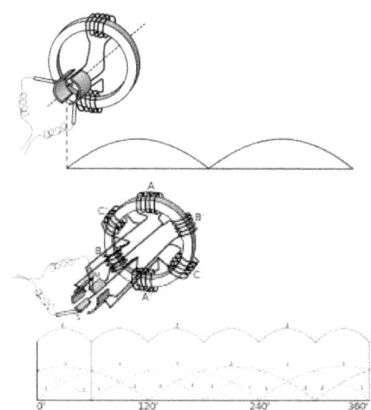

Figure 65: A one pole, two coil Gramme ring (above) and a three pole, six coil Gramme ring (below)

Source: Hawkins Electrical Guide (1914).

Step from permanent magnets to electromagnets: As the strength of the magnetic field is important, and since natural magnetism in a "permanent magnet" is low, increasing this magnetism was a logic step. Originally, the (weak) residual magnetic field of iron was used, but after developing the electromagnet, it was easy to apply this principle to the motor. The iron of the permanent magnet was surrounded by a coil, creating an electromagnet (see Figure 66). Cooke and Wheatstone (and others) developed a dynamo that was constructed with electromagnets rather than permanent magnets.

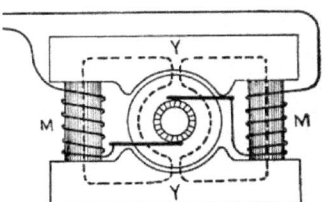

Figure 66: The permanent magnet replaced by the electromagnet called field coil

Source: Hawkins Electrical Guide (1914).

---

and its publication sometimes added several years before the work would arrive in Paris or London."

Step to self-exciting electromagnets:
Another step consisted in taking the current induced in the revolving coil (by the field magnets) and sending it back through the coils around the field magnets (the electromagnets). These currents then created greater magnetism that induced, with an increased efficiency, still stronger currents in the armature coils. The result was much stronger electrical power (see Figure 67). The shunt-wound generator output varies with the current draw, while the magneto output is steady, regardless of load variations. This principle of the "self-exciting" dynamo was the discovery of Sören Hjorth, of Copenhagen, and is fully described in his British patent, № 806 of 1855, for "An Improved Magneto-Electric Battery." Hjorth recognized the advantages obtainable from an electromagnet field system (i.e., that the field strength of the magnetic field may be varied). The same idea was later implemented by Werner Siemens as he also dispensed with permanent magnets and created a generator on the always available residual magnetism of iron.

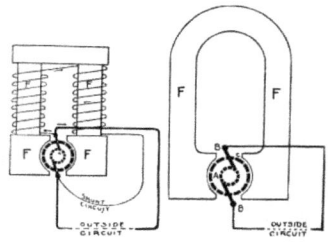

Figure 67: A self-excited shunt-wound DC generator (left) and a magneto DC generator with permanent field magnets (right)

Source: Hawkins Electrical Guide (1914).

Step by step, the fundamental properties of the future electric dynamo were discovered and applied. The electric dynamo would prove to be an important alternative to the battery.

John Woolrich's generator (1844)

One of the early applications for DC electricity was its use by manufacturers to apply an electric current to coat metals: the process of electrochemical plating, in which a thin layer of electro-deposited metals changes the surface properties of another metal. Quite soon after Volta introduced his battery, professor Luini Brugnatell, colleague of Volta at the University of Padua, experimented successfully with this process that resulted in the first electroplating in 1803.

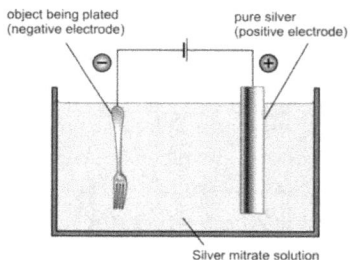

Figure 68: Principle of electroplating

In essence the process of electroplating is the reversal of that of the battery.

In 1836 J. F. Daniell, professor at King's College, London, found during his experiments that metallic copper was deposited on copper cylinders. Moritz Hermann von Jacobi, professor at the University of Tartu, Estonia, repeated Daniell's experiments in 1837 and reported his process of "galvanoplastik" to the Academy of Sciences, published in 1840. It was all about an immersion process resulting in electroplating. The Englishman John Wright (1808–1844) invented the process of electroplating involving potassium cyanide. He sold his knowledge for a total of £1.500 to George R. Elkington (1801–1865), who was active in the electroplating business. Elkington patented his process in 1840: British Patent №. 8.447 of March 25, 1840: "Improvements in Coating, Covering, or Plating certain Metals."

He used electrochemical batteries to create the DC current they needed for the electroplating of gold and silver. Now electroplating became a galvanic process in which electricity was used (Hunt, 1973, pp. 16-23).

Figure 70: Elkington Co. electroplating powered by dynamo generator (1844)

Source: http://myweb.tiscali.co.uk/speel/ otherart/elkingtn.htm

As the usage of wet batteries in the new electrochemical process was quite a costly affair, the engineers and businessmen of that time were highly interested in the newly developed dry battery (a.k.a. magneto-electric generator). It was John Stephen Woolrich who developed a dynamo suitable to supply the electricity for electroplating and installed it in 1844 at the factory of Prime & Sons in Birmingham. He filed for GB-Patent №. 9.431 in August 1842.

Figure 69: Woolrich Magneto generator (1844)

Source: (King, 1962, p. 353)

> *Most of the preceding instruments of the 1830s were essentially laboratory instruments constructed for experimental purposes. In the following decade, John S. Woolrich of Birmingham, England, made one of the earliest commercial applications of magneto generators. In his patent application 9431 of 1841,[53] Woolrich described how Saxton generators could be*

---

[53] British Patent 9431, August 1st, 1842

modified for electroplating, and his method seemed feasible enough to be tried by the Elkington firm in Birmingham, the same English firm that had already pioneered in electroplating (King, 1962, p. 350).

Frederick Holmes's dynamo (1853)

Over time there was a specific application area that stimulated the early development of the dynamo: the lighthouse, located along the coast to guide seagoing traffic. For a long time, lights had been based on natural fuels (such as the Argand lamp). But then the carbon arc light was introduced in the lighthouse (powered by batteries), but these were soon replaced by the world's first generator-powered electric lights. To power these lights, which employed carbon arcs, inventors devised new generators and other components that helped launch the electric-lighting industry. The development of the electric dynamo stimulated the application of arc lights, and the improving arc light made the early adaptation of the electric dynamo possible.

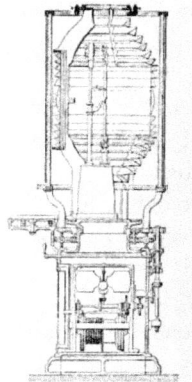

Figure 71: The arc light in the lighthouse at Cap de la Heve (France)
Source: (King, 1962)

From 1859 to the late 1870s, a number of French and British lighthouses became laboratories for long-term testing of arc lights...At any given time, lighthouses of the major maritime powers usually contained a mix of old technologies being retained or phased out, new technologies being phased in, and experimental technologies undergoing trials (Schiffer, 2005, p. 279).

An early example of the lighthouse application of the newly discovered electrical "magneto generator" was found in France and England. In France it was the experiment at the lighthouse at Cap de la Heve that proved the feasibility of arc lighting (Figure 71). In England Frederick H. Holmes demonstrated in 1853 that the dynamo could be used to supply electricity for an arc lamp. It was the first application of generator-powered electrical lights in lighthouses—with their important role in safeguarding those at sea—where the original oil lamps were to be replaced by electrical systems.

For several decades, two technical constraints stymied the commercialization of arc lighting. The first was the heavy drain that carbon arcs put on batteries. Although the invention of so-called constant batteries by the chemist J. F. Daniell and others in the late 1830s ameliorated the problem, even those batteries slowly lost power, and so the light faded. To restore the light's intensity, one had to replace the battery's zinc electrodes and refresh the acid electrolyte, at significant cost. The second constraint was the need to gradually push the carbon rods together as their tips slowly wore down (Schiffer, 2005, p. 282).

Using an electric dynamo instead of batteries could solve one of these problems. Developing more advanced regulators could solve the other. But the battery-arc light combinations never made it to the lighthouse. Everything changed when, in 1856, Holmes was granted a patent for his magneto-electric machine (GB patent № 573 of March 7, 1856).

Figure 72: Holmes electric generator (1857)
Source: (King, 1962)

> *Instead of one disk armature that rotated between the poles of a single bank of permanent magnets, Holmes spun six disk armatures on a common axis between seven parallel banks of permanent magnets. Every other disk was displaced through a small angle so as to reduce the fluctuations of the total induced current* (King, 1962, p. 351).

So the combination of magneto-electric machines (generators) and arc light was realistic. In February 1857, Holmes suggested to Trinity House, a private guild in charge of lighthouses in England, a possible use in lighthouse application for the new electric light system of arc light and generator. He demonstrated it to the Trinity House Light Committee in March 1857 in the latter's experimental "lantern" at Blackwall wharf. Michael Faraday, in his capacity as scientific advisor to Trinity House, had a positive reaction. Trinity House imposed some additional requirements, which caused Holmes to redesign his system—including adaptations such as the steam engine that was directly coupled to the generator and the permanent magnets and coils that were used.

So Holmes started working on a new machine. He was granted a second patent (British patent 2.628 April 14, 1858) for his new generator design. Trinity House approved the installation of a Holmes system in the upper lighthouse (the "high light") at South Foreland, on the cliffs near Dover, for trial in December 1858; however, the results were unsatisfactory. After that attempt another trial was held in the lighthouse at Dungeness (also at the Straits of Dover) with Holmes's newly designed regulator for the arc light. But the combination of high costs, an inefficient commutator, frequent mechanical breakdowns, and untrained personnel prohibited further implementation.

A similar experiment took place in France where the French lighthouse authorities, the *Commission des Phares*, installed an Alliance machine with a Serrin arc light in lighthouse La Hève in 1863 along the Atlantic coast.

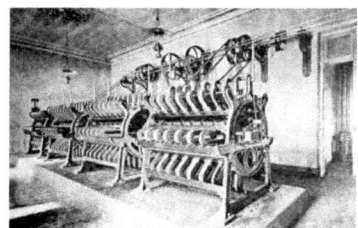

Figure 73: Generator room in the south lighthouse at Cap de la Hève, showing the two sets of Alliance generators
Source: (King, 1962, p. 371)

As no advantages can counterbalance the want of certainty in lighthouse illumination, no further step was taken by the Trinity House in the development of the electric light until the latter part of 1866, when favorable reports were received from the French lighthouse authorities of the satisfactory working of the system at the two fixed lights at Cape La Hève, the south light established in December 1863, and the north light in September 1865 (Douglass, 1879).

Holmes, in the meantime, after hearing of the successful machines developed in France, kept on improving his machines and filed for more patents. He was granted several British patents, but in the end his combination of the magneto-electric engine and the arc light did not work out. The combination did work later, as in 1879 there were ten arc-lit lighthouses in the world. By 1882 there were five electric lighthouses in England and four in France. The idea was good; the realization proved to be difficult. It took decades before the technical problems of the magneto-

Table 2: Some of the British patents granted to F. H. Holmes

Patent №.	Filed	Description
GB 573	March 7 1856	Magneto-electric machine
GB 1.998	July 20,1857	Magneto-electric generator
GB 2.628	October 14, 1857	Magneto-electric generator
GB 2.221	August 1, 1867	Apparatus for producing electric light
GB 2.307	August 10, 1867	Apparatus for producing electric light
GB 2.060	June 26, 1868	Electromagnetic and magneto-electric machines
GB 2.665	August 28, 1868	Electric telegraph/magneto-electric generators
GB 1.744	June 5, 1869	Electromagnetic machines/dynamo-electric generators

Source: (Dredge, 1882), Appendix A: Abstracts of patents. Center for Research Libraries
http://dds.crl.edu/loadStream. asp?iid=17444&f=8;
http://dds.crl.edu/loadStream.asp?iid=17444&f=9 (Accessed November 2014)

generator arc systems were solved. It was only after the discoveries of Werner Siemens and Charles Wheatstone that dynamos could be created with electromagnets instead of permanent magnets (the self-exciting dynamo) and that the dynamo was used more and more in lighthouses.

Sören Hjorth's dynamo (1854)

The Dane Sören Hjorth (1801–1870), a civil engineer who experimented with steam cars and was involved in Denmark's first railway, became interested in Oersted's discovery of the electro mechanism and started to experiment with electricity. His particular interest was in electromotive engines and—what he called— "dry batteries" (as opposed to the expensive electrochemical "wet battery" from Volta).

Figure 74: Hjorth's electromagnetic reciprocating engine (1849)
Source: www.tipsimages.it/

At that time, the use of steam cars on the country roads attracted great attention in England, and many different constructions appeared. In 1834 Hjorth, aided by subventions from the Rejersen Foundation and the Government, went to England, in order to acquaint himself with the use of these steam cars on highroads and railroads. During these years he very actively investigated the use of steam power, especially as a means of propulsion for vehicles and ships. With admirable interest and diligence, he studied the steam-propelled road carriage, and for a long time he considered that to be the future means of conveyance. Although he did not succeed in getting his own steam carriage put to practical use, he made many experiments on a steam car, and I am told by one of his passengers that on the level streets of Copenhagen and Frederiksberg all went very well, but the carriage could not climb Valby hill (Smith, 1912, p. 3).

His road-based steam-carriage experiments were not that successful, and his interest was attracted to the use of rails for steam carriages, a thing that was going to be the future. After an involvement in the franchise for a railroad company in Denmark (the Stockholm-Roskilde-seaport line) with a stock

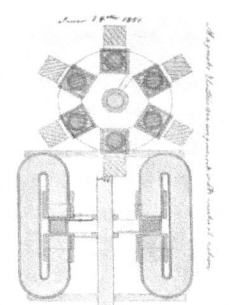

Figure 75: Sketchbook drawing of Hjorth magneto-electric engine (1851)
Source: (Smith, 1912)

capital of 1.5 million "rixdollars" (a very considerable sum in those days), he became director of the Sealand Railroad Company (1844). The money was raised, the line was built and did open in 1847. So, it is safe to conclude that the steam engine and its application played an important part in his early life (Smith, 1912, p. 6).

As early as in 1842, Hjorth started experimenting with electromagnetic machines. He went to England in 1848 to have a prototype made of his idea for a reciprocating machine running on voltaic batteries. (See Figure 74: Hjorth's electromagnetic reciprocating engine (1849).) To pay for the development of the machine, he created a partnership with some financial investors and managed to get Danish patent number 12.295 granted on April 26, 1849. His machine was shown at the Royal Society, the annual Meeting of the Society of Civil Engineers in 1851, and the Universal Exhibition in London in 1851 (Smith, 1912, pp. 8-11). The problem with his first design for an electromagnetic machine, which even looked like a steam machine, was the energy supply from the—in those days expensive—wet batteries. It was obvious to Hjorth that he needed an alternative source of electric supply—a "dry battery." He started experimenting again, and in his sketchbook he noted on May 1, 1851:

Figure 76: Hjorth dynamo (1855)
Source: (Byrn, 1900)

> *By passing the current on the said way round the Electromagnets, these will of course be excited in proportion to the strength of the same, and the more they are excited, the more will the discs be influenced by the magnets, a mutual action thus taking place* (Smith, 1912, p. 12).

In October 1854 Hjorth was granted Danish Patent №. 2.198 for an electric dynamo. He used permanent magnets that were coiled to become electromagnets. The dynamo that resulted from these efforts was patented in England in 1855 (GBK patent 806). Hjorth describes the action of the battery (i.e., dynamo) as follows:

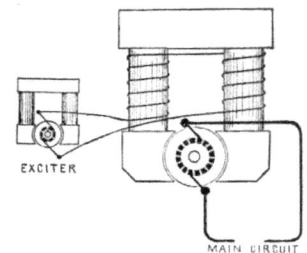

Figure 77: Principle of self-exciting dynamo with a separate generator feeding the electromagnets
Source: Wikipedia Commons

> *The permanent magnets acting on the armatures, brought in succession between their poles, induce a current in, the coils of the armatures, which current, after having*

been caused by the commutator to flow in one direction, passes round the electromagnets, charging the same and acting on the armatures. By the mutual action between the electromagnets and the armatures, an accelerating force is obtained, which in the result produces electricity greater in quantity and intensity than has heretofore been obtained by any similar means (Smith, 1912, p. 18).

He was describing the self-exciting magneto generator. This principle is called "self-exciting" because, instead of using the magnetic field created by the permanent magnet, the—much stronger—magnetic field created by an electromagnet is used.

Soon, Hjorth also patented reciprocating and rotary electric motors (UK patents 807 and 808, 1855). But he did not succeed in commercializing his invention. In 1866 Henry Wilde published a paper about his machine, in which a second engine produced the current needed to magnetize the electromagnets. This is exactly the same principle underlying the dynamos built by Hjorth in 1854 and 1855.

Table 3: Some of the British patents granted to Soren Hjorth

Patent №	Filed	Description
GB 12.295	October 26, 1848	Applications of electromagnetism as motive power
GB 2.198	October 14, 1854	Magneto-electric battery/Dynamo-electric generator
GB 2.199	October 14, 1854	Electromagnetic machine/Electromagnetic motor
GB 806	April 11, 1855	Dynamo-electric generator/Auto-exciting
GB 807	April 11, 1855	Electromagnetic machine
GB 808	April 11, 1855	Electromagnetic engine

Source: (Dredge, 1882), Appendix A: Abstracts of patents.. Center for Research Libraries http://dds.crl.edu/loadStream.asp?iid=17444&f=8 (Accessed November 2014)

Henry Wilde's dynamo (1863)

Henry Wilde (1833–1919), a mechanical engineer, established a business in Manchester, England, as an electric-telegraph and lighting engineer in 1856. With his brother-in-law, a silversmith, he created a company called Wilde & Co. Responding to the business opportunities offered by the fast expansion of telegraphy, he developed the "ABC-telegraph system," rivaling Charles Wheatstone's system. For the sales division of this system, he created the Globe Telegraph Company.

The chief object of the company was to establish "a system of private telegraphic communication between Public Offices, Police, Fire and Railway Stations, Banks, Docks, Mines, Manufactories, Merchants' Offices, etc." For the creation of such a company, in accordance to the Telegraph Act of 1863, a permission was needed; a so-called *"Act of Parliament,"* which he obtained, at considerable costs, in 1864. However, some years later the Telegraph Act of 1868 gave the government a monopoly on telegraphy. Against this Wilde petitioned, and in his evidence before a Select Committee, he claimed that his patent rights would be greatly depreciated, if not entirely destroyed, by the Act. Much to his disappointment, the Committee decided against his claims, and his Globe Telegraph Company ceased its business (Haldane Gee, 1920, p. 2).

A consequence of Wilde's telegraphy work had been his experiments with producing electric currents. These experiments resulted in a series of patents (among those GB patent №. 3006 of December 1863, in which he describes his dynamo generator, with its separate permanent-magnet exciter mounted above and driven from a common shaft). For this discovery he also obtained US patent №. 59.738 on November 13, 1866. In the US patent, he indicated that "This invention relates to certain improvements in that class of machines known as 'magneto-electric machines,' which improvements are also applicable to electromechanical machines." In other words the self-exciting machine could either be an electric dynamo creating electricity, or an electromotive motor creating rotative power. However, these electromagnetic generators had a very serious defect. The magnetic currents in the armature, being converted into heat, produced a rise of temperature of three hundred degrees Fahrenheit and upwards.

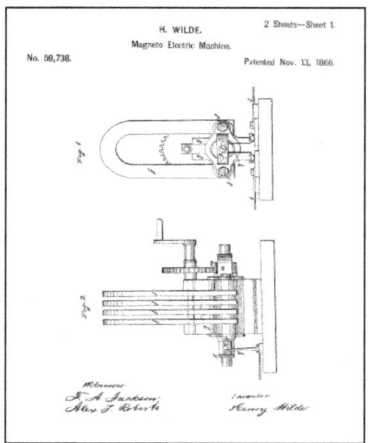

Figure 78: Henry Wilde's US patent No 59.738 (Nov. 13, 1866)
Source: USPTO

Wilde introduced the principle of *accumulation by successive action*, by combining two of these cylindrical armature machines, one larger than the other—the larger machine being furnished with electromagnets instead of with permanent magnets. In his paper "Experimental Researches in Magnetism and Electricity Part I" of April 26, 1866, he wrote the following:

> *But to direct attention to some new and paradoxical phenomena arising out of FARADAY'S important discovery of magneto-electric induction, the close consideration of which has resulted in the discovery of a means of producing dynamic electricity in quantities unattainable by any apparatus hitherto*

constructed…for I have found that an indefinitely small amount of magnetism, or of dynamic electricity, is capable of inducing an indefinitely large amount of magnetism. And again, that an indefinitely small amount of dynamic electricity, or of magnetism, is capable of evolving an indefinitely large amount of dynamic electricity (Wilde, 1867, p. 90).

He also describes the construction of an "apparatus" that "was of the same form as that used by Siemens," with which he experimented. One of those experiments was with electromagnets that were powered by the machine itself:

Figure 79: Henry Wilde's generator of 1866
Source: www.gutenberg.org/files/44502/44502-h/44502-h.htm#i_043

A second series of experiments was made with the view of ascertaining the relation existing between the lifting power of the permanent magnets on the magnet-cylinder, and that of an electromagnet excited by the electricity derived from the magneto-electric machine…it appeared reasonable to suppose that a large electromagnet excited by means of a small magneto-electric machine could, by suitable arrangements, be made instrumental in evolving a proportionately large quantity of dynamic electricity, notwithstanding the pulsatory character of the electricity transmitted through the wires surrounding the electromagnet…A comparison of the heating effects of the two machines, as found by these experiments, brings out the important fact, that a much greater amount of electricity is evolved from the electromagnetic machine than is evolved simultaneously from the magneto-electric machine (Wilde, 1867, pp. 93-103).

So, using a small additional machine (see Figure 79) to power electromagnets of the main machine, he developed the predecessor to the "self-exciting" dynamo. His paper describing the experiment, "On a new and powerful Generator of Dynamic Electricity," aroused considerable attention.

In 1867 he obtained a patent for a totally different design for a machine used for arc lighting and electrodeposition. Henry Wilde had thus produced two commercial types of generators, which could be used to replace the primary batteries used in electrochemistry and for arc lighting. His generators were used by the electroplating firm of Elkington, from which he received three hundred pounds as royalties during a number of years.

In 1877 Henry Wilde petitioned the Privy Council to extend his dynamo patents of 1863 and 1867. He had the advantage of the evidence of the eminent engineer, Y.J. Bramwell, who made the claims of the patentee very clear. The result of the petition was that the patents were extended until 1884 (Haldane Gee, 1920, p. 6).

Wilde had to defend his patents, without much success. But it gave him experience in litigation.

When the Gramme dynamo was introduced into this country, Wilde brought an action against the British agents for infringement of his patent. The agents obtained the opinions of F. H. Holmes, who for twenty-five years had been engaged in designing and constructing dynamos, S. A. Varley, one of the first to use residual magnetism for the excitation of electromagnets, and Fontaine and Werdemann, well-known inventors. Their evidence was so strong, and threw so much doubt on the priority of Wilde's inventions, that Wilde found it advisable to withdraw his action (Haldane Gee, 1920).

As Wilde had introduced the principle of *accumulation by successive action*, by combining two of these cylindrical armature machines (the larger machine being furnished with an electromagnet instead of with permanent magnets), he wanted recognition as being the inventor of the "dynamo-electric machine." He decided to sue S. P. Thompson who in his eyes, in his publication *Dynamo-electric Machinery* (Thompson, 1896), did not give him enough recognition as inventor:

Table 4: Some of the British patents granted to Henry Wilde

Patent №	Filed	Description
GB 858	April 1861	Electromagnetic telegraphs/Magneto-electric generator
GB 1.994	July–September 1861	Electromagnetic telegraphs/Magneto-electric generator
GB 2.997	November 1861	Electromagnetic telegraph/Magneto-electric generator
GB 2.845	October 22, 1862	Electromagnetic telegraphs/Magneto-electric generator
GB 3.240	December 3, 1862	Electromagnetic telegraphs/Magneto-electric generator
GB 516	April 1863	Electromagnetic telegraphs/Magneto-electric generator
GB 3.006	December 1,1863	Electric telegraph/Magneto-electric motor
GB 1.412	May 23, 1865	Producing and applying electricity
GB 2.762	October 1865	Electromagnetic telegraphs/Magneto-electric generator
GB 3.209	December 4, 1866	Electromagnetic and magneto-electric machines
GB 842	March 23, 1867	Electromagnetic and magneto-electric induction machines

Source: (Dredge, 1882), Appendix A: Abstracts of patents. Center for Research Libraries http://dds.crl.edu/loadStream.asp?iid=17444&f=8 (Accessed November 2014)

> *With the aim of restoring his supposedly damaged status as sole inventor of the dynamo…Wilde sued Thompson for libel. The occasions for this were Thompson's publication* The life of Faraday *and the preparation of the sixth edition of the textbook* Dynamo-Electric Machinery. *In both cases Wilde felt that Thompson granted him insufficient credit for his contributions to the invention of the "self-excited" electric generator*
>
> (Arapostathis & Gooday, 2013, pp. 114-115).

Five years earlier Wilde had already started a campaign for the restoration of his reputation in the scientific and engineering community. In 1898 he wrote a letter to the Institution of Civil Engineers, of which he was an honorary member:

> *During the course of his experiments with the later form of his telegraph instruments, your Honorary Member observed the great increase of power of an electro magnet above that of the permanents magnet of the magneto-electric machine which exited it. This observation led him up to the discovery of the principle and invention of the dynamo, as the machine is now called. The invention is described and figured in the specifications of his patents of December 1863 and October 1865* (Arapostathis & Gooday, 2013, p. 115).

It became a semantic issue, in a way. What was the meaning of the word "dynamo" (as introduced by Charles Brooke) versus the "dynamo-electric machine" (as introduced by Siemens)? Wilde wrote, in a series of letters, that he claimed the word was used to denote the machine of his invention (Wilde, 1900). He addressed these letters to the Society of Arts, who intended to award him the Albert Medal in 1900—to be presented by the king. Wilde also protested the fact that the awarding of the medal did not mention him as being the inventor of the dynamo-electric machine: "No mention is made of my invention of the dynamo-electric machine and its successful application by me to electric lighting and to the electrodeposition of metals from their solutions" (*Ibidem*, Preface). The Society responded with a proposal of a change in the text that did not satisfy Wilde. He then indicated that he was not inclined to accept the medal:

> *Considering the peculiar circumstances under which the partial recognition of my claims has been made by the Council, it will hardly be expected that I can accept the honour conferred upon me with that feeling of gratitude which such an award would naturally call forth. I will, therefore, at some future time, let the Council know when it will be agreeable to me to receive the Albert Medal* (Ibidem).

The Society was not pleased and told Wilde so. The secretary wrote to him:

> *I am instructed by the President and the Council of the Society of Arts to inform you, with reference to your letter of the 12th inst., that they are much surprised and dissatisfied at the manner in which the award of the Albert Medal has been received by you. They further desire me to say that, inasmuch as the last sentence of your letter seems to leave it uncertain whether you propose to receive the medal or not, they will be obliged to you if you will state definitely whether or not you do wish to receive it. If you do, I have instructions to send it to you at once* (Ibidem).

Wilde indicated that he was willing to receive the award by sending a telegram on July 30: "Letter received. Acceptance of the Albert Medal by me fully implied in my correspondence with the Society, and is now confirmed." So the Society *mailed* him the award on July 31. This resulted in Wilde's reaction on October 21:

> *Considering that the presentation of the Albert Medal is usually attended with some amount of ceremony, and has not, for some years past, been made to British recipients before the year following the award, the posting of the Medal to me on this occasion is obviously not intended as a compliment, but is a further manifestation of the same spirit that attempted to thrust upon me the award of a medal on terms which I disapproved of…The service the Society of Arts has thus incidentally rendered to philosophy, and indirectly to religion, by the recognition of the incommensurableness of the electric and magnetic forces, may, at some future time, obliterate the sense of wrong that has been done me by the Council in ignoring my inventions and their applications by me to the Arts* (Ibidem).

On October 25, 1900, he returned the medal, stating, "I enclose herewith the Medal as a contribution to the collection [to the museum of the Society]." All this correspondence, the semantic dispute, and the legal battle was not so much about money, as much as about the recognition of Wilde's scientific and engineering status. Among his peers Wilde lost a lot of goodwill, as well as losing the legal case against Thompson:

> *It was brought before Mr. Justice Buckley in the Chancery Division of the High Court of Justice, who ordered that the statement of claim be struck out on the ground that it disclosed no reasonable basis of action. The action was dismissed with costs against the plaintiff. Against this decision Henry Wilde appealed, but without success* (Haldane Gee, 1920, p. 7).

Moses Farmer's dynamo (1865)

Moses G. Farmer (1820–1893), an American electrical engineer working at a telegraph company, experimented in 1845 with electric devices such as the telegraph, batteries, and electric motors. Farmer gave up teaching, and in 1847 he became a wire examiner for a telegraph line between Boston and Worcester. While on this job, he studied telegraphy, and the following year he was made an operator at Salem, Massachusetts. About this time he began the work that led to his invention of an electric fire-alarm system. In 1851, with William F. Channing, he installed his system in Boston, the first such system in the United States. Farmer became its superintendent. He resigned this post in 1853 and for many years held a succession of jobs, not all of which were connected with electricity.[54]

In 1857 Farmer received US patent № 17.355 for an electric fire-alarm system (the fire-alarm telegraph) for cities. In 1858–1859 he produced electric lamps, and in 1868, with a dynamo of his own invention, he illuminated a room of his house (in Salem) for several months—the first known case of domestic incandescent lighting (Abernathy & Clark, 1985). Edison used one of Farmer's dynamos to power his experimental light bulbs (which were of a completely different design). In 1872 Farmer was appointed to the office of electrician at the United States Torpedo Station at Newport, Rhode Island, as a professor of electrical science. It was Farmer who wrote a letter in 1866 to Henry Wilde that was read at the meeting of the Manchester Literary and Philosophical Society in February 1867:

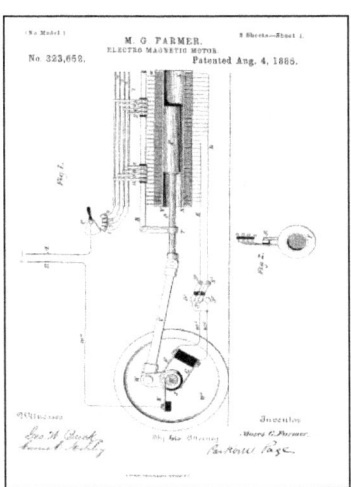

Figure 80: Moses G. Farmer's US patent No. 323.652 (August 4, 1885) for an electric reciprocal motor

Source: USPTO

> *I have built a small machine in which a current from the thermo battery excites the electromagnet of your machine to start it, and after the machine is in action, a branch from the current of the magneto (i.e., armature) passes through the electromagnet, and this supplies the magnetism required*
> (MacKechnie Jarvis, 1955b, p. 569).

---

[54] Text from: http: Moses Gerrish Farmer Facts //www.yourdictionary.com/moses-gerrish-farmer#biography. (Accessed December 2014)

Farmer was a pioneer of many aspects of nineteenth-century electrical invention, but, because he and his wife were spiritualists, they felt that their talents were God-given, and he felt that they shouldn't take credit for any of his inventions. He was granted several patents for his diverse work— including the generator he built together with the Wallace brothers of Ansonia, Connecticut.

Table 5: Some of the patents granted to Moses G. Farmer

Patent №	Granted	Description
US 8.920	May 4, 1852	Electromagnetic bell: machine for striking bells by electromagnetism
US 15.373	July 22, 1856	Improvement in self-acting electric telegraph: alphabetic keyboard design and printing letter wheel
US 17.355	May 19, 1857	Improvement in electric magnetic fire-alarm telegraphs for cities
US 25.003	August 9, 1859	Windlass: winding apparatus for raising and lowering heavy weights
US 72.616	December 2, 1867	Improvement in lighting and extinguishing gas: electrical spark for igniting a gas lamp
US 109.603	November 29, 1870	Improvement in thermoelectric battery: for a battery that could be heated
US 126.628	May 14, 1872	Improvement in electromagnetic machines: construction to prevent the induction spark on the commutator
US 148.289	March 10, 1874	Improvement in apparatus for firing fuses by electricity: a safety device to ignite fuses
US 161.874	April 13, 1875	Improvement in magneto-electric machines: using rotating electromagnets
US 213.643	March 25, 1879	Improvement in electric lights: electric current passing sticks of carbon in a globe with an artificial atmosphere
US 320.234	June 16, 1885	Regulator for dynamo/electric machines
US 322.169	July 14, 1885	Apparatus for refining copper by electricity: equipment for electroplating
US 323.652	August 4, 1885	Electric magnetic motor: reciprocating electric motor with crank
US 460.572	October 6, 1891	Printing telegraph: telegraph that printed letters on a paper strip by a type wheel

Source: USPTO

The self-exciting electric dynamo

As outlined above, some efforts had already been made in the period preceding 1866 to apply the principle of self-excitation by people such as Sören Hjorth, Wilhelm Sinsteden, Henry Wilde, and Moses Farmer. However, they did not lead to considerable results.

But then came the nearly simultaneous developments realized by Varley, Siemens, and Wheatston: the self-exciting dynamo.[55] Werner Siemens applied the distinctive term "dynamo-electric machine," in contrast to the usual term "magneto-electric machine," to this new kind of generator in his announcement of the new principle. Since then, the term has been shortened to "dynamo" (King, 1962, p. 379).

Samuel Varley's dynamo (1866)

It was the Englishman Samuel Alfred Varley (1828–1883)—brother of the engineer Cromwell Fleetwood Varley (1828–1883) and member of a family of artists and engineers—who worked on the self-exciting dynamo. Samuel Varley and Michael Faraday were also distantly related and were both members of the same church (the Sandemanian Church).

> *When a small child I played with magnets and electrical apparatus, and I received many a kindly pat on the head from Michael Faraday, who was an elder of a small religious sect to which my father belonged. I grew up with so deep a reverence for Faraday, both as a religious exemplar as well as the greatest of scientists, that it bordered almost on fear. In the year 1846 at the age of fourteen, I commenced researches to try and ascertain the nature of magnetism...and step-by-step I was led up to the discovery of what Lord Kelvin has aptly termed the "dynamo principle"* (Jeffery, 1997, p. 269).

In 1861 Samuel Varley left his job as an engineer at the Electric and International Telegraph Co. to take over the management of a telegraph-manufacturing business. There he developed a self-exiting dynamo that was completed in 1866—an important discovery that resulted in conflict later when the credit for the invention was claimed and disputed by others. In December 1866 Cornelius and Samuel Alfred Varley (father and son) filed an application for a British patent (№. 3.394 of 1866) under the title "Improvements in the Means and Apparatus for Generating Electricity" (Figure 81, left). The specification describes a self-excited electromagnetic generator in which the dependence of the field system upon the residual magnetism in building up the field is clearly recognized.

> *We construct our apparatus as follows: We wrap soft iron bars with insulated wire in a similar way to an ordinary electromagnet; these bars may be U-shaped, and become electromagnets when the apparatus is in use; we also construct iron bobbins of such a length that they will pass just freely between the poles of the*

---

[55] A self-exciting dynamo is a dynamo with electromagnets that are powered by the current created by the dynamo itself. This is in contrast with the permanent magnets that were used to create a magnetic field.

electromagnets and wrap them with insulated wire (Dredge, 1882, pp. Abstracts of Patents, CXXI).

As Varley's specification was not published until July 1867, it seems that Wilde's description of a self-exciting machine he had made precedes Varley's discovery. The invention of the self-exciting principle was related to his later efforts in the compound-wound dynamo that he patented.[56]

> *The dynamo, thus started on its course, proved to be full of possibilities for the electrical industry, and Alfred worked unremittingly at the problem of improving it. Ten years after his discovery of the dynamo principle, the idea of regulation came to him, and the result was the invention of the compound-wound machine in 1876. Unfortunately he allowed this patent to lapse in 1878*
> (Lee, 1932, p. 963).

In Table 6 some of the patents issued to Samuel Alfred Varley, in cooperation with his brothers, Octavius Varley and Frederick Henry Varley, are shown.

Table 6: Some of the British patents granted to Samuel A. Varley (1), O. Varley (2), and F. H. Varley (3)

Patent №	Filed	Description
GB 3.394	December 24, 1866	Generating electricity/Dynamo-electric generator (1+2)
GB 1.755	June 15, 1867	Electric telegraphs/Dynamo-electric generator (1+2)
GB 3.329	October 31, 1868	Generating static electricity/Electrical generator (1)
GB 2.525	August 25, 1869	Transmitting and recording electric signals/ Electromagnetic generator (2+3)
GB 131	January 18, 1871	Telegraph apparatus/Magneto-electric generator (1)
GB 1.150	April 29, 1871	Electric telegraph apparatus/Magneto-electric generator (1)
GB 4.905	? 1876	Compound-wound dynamo

Source: (Dredge, 1882), Appendix A: Abstracts of patents. Center for Research Libraries, http://dds.crl.edu/ loadStream.asp?iid=17444&f=8, http://dds.crl.edu/loadStream.asp?iid=17444&f=9 (Accessed December 2014)

---

[56] There are several methods to feed the electromagnet with a current. Basically it can be in series (series-wound dynamo) or in parallel with the load (the shunt-wound dynamo). In a compound-wound dynamo, the supply for the electromagnet comes in part from the serial connection and in part from the parallel connection with the load.

In the case of the compound-wound dynamo, Varley filed GB patent № 4.905 in 1876 for an innovative form of dynamo that featured a high and well-regulated terminal voltage. However, Brush had been granted a patent for a compound-wound dynamo in 1878 (filed by his UK patent agent Herbert John Haddan), and on this ground his company tried to suppress its competitors in the industry (Figure 81, right). Interestingly, these were just two of the 171 patents that were issued between 1877 and 1885 relating to the design of the compound-wound dynamo.

> *The rapid exploitation of [incandescent] lighting created a demand for large, efficient, self-regulating dynamos. Then in 1878 the Brush Company of America took the initiative. Only ten months after Varley's unprotected patent had been published, they filed another patent describing the compound-wound machine in almost the same words as Varley's 1876 patent. The Anglo-Brush Corporation then ostentatiously purchased American Brush's patent for £100.000 and set about pursuing a number of English firms who had already taken up the manufacture of Varley's machine, demanding large royalties from them. A syndicate of manufacturers was formed to fight these demands, and Varley was retained to dispute the technical aspects of the claims made by Brush. The case of Brush v. Crompton was scheduled for December 1887, but on the eve of the trial, the syndicate, headed by Crompton, considered themselves to have been outmanoeuvered by Brush and capitulated. What happened was that the Brush Corporation had been astute in retaining the services of expensive barristers, patent experts, and eminent scientists, people whom Varley chose to call "that trades unionism of special pleading experts." They frightened so many in the syndicate that they came to the conclusion that discretion was the better part of valour* (Jeffery, 1997, p. 272).

Figure 81: Varley's dynamo (1866, left) and compound-wound machine (1876, right)
Source: (Jeffery, 1997, pp. 270, 273)

The situation looked like a lost case. However, in a lawsuit that followed in the Court of Session, Edinburgh, in November 1888, judgment was given in favor of Varley's precedence in the invention of the compound-wound machine, the Brush patent being declared invalid.

> *A Scottish firm, after conferring with Varley, came to the conclusion on commercial grounds that it would be in their interest to resist Brush, and a new strategy was adopted. Instead of waiting to be attacked, the firm started proceedings in Scotland accusing Brush of damaging their business by threatening their customers. This completely wrong-footed Brush because their highly competent and expensive lawyers were not permitted to practise in the Scottish Courts. They tried everything to get the case heard in England, but failed. They then made a monetary offer to the Scottish firm that was so substantial that their legal advisers recommended acceptance. A compromise solution was reached. The Scottish firm, King Brown and Co., agreed to continue with the action provided a guarantee fund could be raised. Varley, by issuing a pamphlet and by canvassing all leading dynamo manufacturers, was successful in raising sufficient pledges. In the pamphlet he was able to refute the arguments already put forward by the experts in the English court action. For example, Sir William Thomson, in his affidavit submitted on behalf of Brush, speciously asserted: "Varley has not disclosed the idea of a stronger magnetic field [provided by the winding] when external work is being done, and no external work is being done," a statement that Varley maintained "went without saying." He also pointed out that, in any case, neither had Brush made this claim in its patent. The case was heard at the Court of Session, Edinburgh, in 1888, and Brush was defeated by the strength of Varley's position. Brush subsequently appealed to the Inner House of the Court and to the House of Lords, but both appeals were disallowed. Nevertheless, although he was now acknowledged to be the inventor of the compound machine, he received no pecuniary benefit* (Jeffery, 1997, p. 274).

Samuel Varley died in 1921, a poor man.

Werner von Siemens's dynamo (1866)

The German Ernst Werner Siemens (1816–1892) was born into a middle-class family with fourteen children. His parents were tenants who suffered from the agricultural crisis of 1818–1825. He went to the Gymnasium in Lübeck but could not get much further schooling due to financial problems. So he joined the Prussian army (the Prussian Engineering Corps) in 1834. After that he was educated at the Prussian Military Academy's School of Artillery and Engineering in 1835. This training created the basis for his later engineering activities. (His military duties left him obviously time for scientific experiments with electrolytic plating with silver and gold.) In 1842 Werner was awarded his first patent, on a gold-plating process. The patent was sold to the English firm Elkington & Co for £1600. [57] This created a more solid financial base in Siemens's life. He then pursued the more interesting and emerging field of electricity that, after Oersted's and Faraday's discoveries, was just emerging and creating a lot of industrial activity.

Fast, reliable communication was of special interest to the army, and Siemens concentrated his efforts on providing it. Having observed an early version of an electrical telegraph, the budding engineer realized the device would need to be vastly improved to meet the needs of the army. By 1847 Siemens had built his own version of the telegraph that was significantly superior to any previously constructed. This telegraph was built by the mechanic Johann George Halske (1814–1890). In 1847 Siemens and Halske formed the company *Telegraphen-Bauanstalt Siemens & Halske*. Siemens then built a telegraph network for the Prussian government (Berlin-Cologne, Berlin-Frankfurt, completed in 1849). After resigning from the army in 1849, he concentrated on his company. Then he expanded internationally, as one of Siemens's brothers represented him in England (Wilhelm, later Sir William Siemens) and another in St. Petersburg, Russia (Carl Siemens). He got a contract for a Russian telegraph line from St. Petersburg to Moscow and the Crimean peninsula in 1851. Later his brother Carl managed the Russian subsidiary that expanded during the Crimean War (1853–1856). Wilhelm managed the London subsidiary that later became the independent Siemens, Halske & Co. (Feldenkirchen, 1994).[58]

Werner Siemens continued experimenting with dynamos, and in 1854 he introduced an important improvement—now known as the Siemens double-T armature winding. This dynamo included the adoption of an electromagnet of peculiar form: the H-armature. Siemens used this armature in the first dynamo he built in 1866 (Figure 82). In that same year, Werner Siemens realized that the residual magnetism in the iron in

Figure 82: Siemens's self-exciting dynamo (1866)

Source: Deutches Museum, Munich

---

[57] This would be a value of £1,078,000 in 2010, using the labor value calculation. Source: (www.measuringworth.com)
[58] See: B.J.G.van der Kooij: *The Invention of the communication Engines*. (2015)

a dynamo was sufficient to induce a weak voltage in the rotating armature. The resulting current can be used to continually boost the magnetism to the saturation point. This "self-excitement" was termed the *dynamo*-electric principle and it replaced the *magneto*-electric principle. Siemens created the "H" or shuttle armature that was patented (British patent 261) in England in September 1867 by his brother Charles W. Siemens (1823–1883).[59]

> *The Siemens dynamo was in all respects an immense leap forward and as opposed to the Wilde dynamo it was marketable. It could produce a steady and stable current, did not demand too much maintenance, and was easy to implement in a factory production, which consequently resulted in a fierce competition to copy the Siemens dynamo: "The world's most efficient dynamo with respect to producing electric current was the one that E. Werner and C. William Siemens, brothers working in Germany and England, respectively had patented"* (McPartland, 2006, p. 122).
>
> *So, in 1866 Werner Siemens discovered the dynamo-electric principle of the self-exciting electricity generator and submitted a paper to the Berlin Academy of Sciences for reading on January 17, 1867. It was published some months later, and his conclusions were communicated to the Royal Society of London in a paper written by his brother, who exhibited a hand-driven generator on February 14, 1867. At the same meeting, another pioneer, Professor (afterwards Sir) Charles Wheatstone, FRS (1802–1875) read a paper on the same subject and also exhibited a hand-driven generator similar to that of Siemens except in respect of the winding details* (MacKechnie Jarvis, 1955a, p. 568).

Thus it seems that Dr. Werner Siemens and Sir Charles Wheatstone discovered this type of self-exciting machine almost, if not quite, simultaneously. Abandoning the use of an additional magneto-electric machine, as used by Wilde, and without invoking a succession of electromagnets, they succeeded in generating from perfectly infinitesimal beginnings enormous amounts of electric power.

As Germany did not have a patent system in those years (just after 1877), Siemens could not apply for German patents. Siemens's brother William, located in England, was granted a patent for his dynamo (British Patent №. 261 of 1867). Later Werner Siemens was granted a British and American patent for his dynamo-electric machine (GB 760 of February 16,

---

[59] As it was not until May 25, 1877, that the German Patent Act (Patentgesets) was created and not until July 1877 that the German Patent Office (Kaiserliches Patentamt) was established, patenting was not possible in Germany. There existed however a Prussian patent.

1882, and US patent №. 264.780 of September 19, 1882). He also obtained other patents, but they were for insulating conductors (US patent №. 395.083 of December 25, 1888), an electric meter (US patent №. 415.557 of November 19, 1889), an electroplating process (US patent №. 460.354 of September 29, 1891), and a concept of an electric railway (US patent №. 322.859 of July 21, 1885, US patent №. 520.274 of May 22, 1894).

Charles Wheatstone's dynamo (1866)

The Englishman Charles Wheatstone (1802–1975) was apprenticed at the age of fourteen to his uncle, a maker and seller of musical instruments. After his uncle's death in 1823, he took over the business along with his brother. As he was more interested in science than in business, he experimented with the transmission of sound. His results were developed into public exhibitions in London. One of these was the Enchanted Lyre that was shown in various places for a couple of years.

Figure 83: Wheatstone's dynamo (1866)
Source: (King, 1962), p.379

> *In May of 1823 the Danish scientist Hans Christian Oersted (1777–1851) visited London and saw the Enchanted Lyre. Wheatstone was not then acquainted with the scientific community, and Oersted provided his introduction. Wheatstone and Oersted found that they had in fact performed several similar experiments, and Oersted encouraged Wheatstone to write his first scientific paper, which was read at the Academy of Sciences in Paris in June 1823, and published the same year in England* (Bowers, 1975, p. 502).

Wheatstone continued his experiments with sound after he became a close friend of Michael Faraday. His most significant practical work in sound was the development of the "concertina," an instrument that was manufactured by his firm. In 1834 Wheatstone, who had won a name for himself, was appointed to the Chair of Experimental Physics in King's College, London. After some experimenting with spectroscopy, he lectured, in the mid-1830s, on a system by Baron Schilling (the five-needle telegraph) and became excited by the idea of telegraphy.

> *In early 1837 he met with William Fothergill Cooke (1806–1879), a surgeon and professor of anatomy at the University of Durham, who had devised a telegraphy system of his own. Cook had in 1846, with the businessman and politician Lewis Ricardo (1812–1862), created a company called the Electric*

Telegraph Company. In May 1837 they agreed to join their forces: Wheatstone contributing the scientific, and Cooke the business affairs. The deed of partnership was dated 19 November 1837. They together developed a (two needle) telegraph for which a UK patent was filed in May 1837, and granted on 12 June 1837. It was the first English patent for an electric telegraph. This patent was followed by a Scottish patent that was sealed on December 12th 1837 (Bowers, 2001, pp. 119-129).

Charles Wheatstone had, in the early 1840s, made a series of electric motors to which he gave the name "eccentric electromagnetic engines." As he needed an alternative energy source for his battery-powered telegraph system, he became interested in the electric dynamo. After being appointed professor of experimental physics at Kings College, London, he worked on the telegraph and the electric dynamo. He presented his findings on the self-excited (shunt-wound) dynamo to the Royal Society on February 1867 (Figure 83). The Siemens Dynamo had a similar design, with the difference being that in the Siemens design, the stator electromagnets were in series with the rotor, but in Wheatstone's design they were in parallel. Some of the British patents granted to Wheatstone are shown in Table 7.

Table 7: Some of the British patents granted to Charles Wheatstone

Patent №	Filed	Description
GB 7.390	June 12, 1837	Electric telegraph/Conductors (with W. F. Cooke)
GB 8.345	January 21, 1840	Electric telegraph/Magneto-electric generator
GB 9.022	July 7, 1841	Producing, regulating, and applying electric currents/Magneto-electric generator
GB 10.655	May 6, 1845	Electric telegraphs/Electromagnetic generator
GB 1.241	June 2, 1858	Electromagnetic telegraphs/Electric motor
GB 2.462	October 1860	Electromagnetic telegraphs/ Magneto-electric generator
GB 2.172	August 18, 1871	Telegraphs/Magneto-electric generator
GB 473	February 15, 1872	Electromagnetic telegraphs/Magneto-electric generator

Source: (Dredge, 1882), Appendix A: Abstracts of patents. Center for Research Libraries, http://dds.crl.edu/ loadStream.asp?iid=17444&f=8; http://dds.crl.edu/loadStream.asp?iid=17444&f=9 (Accessed December 2014)

Zenobe Gramme's dynamo (1871)

Zenobe Gramme (1826–1901) was a Belgian carpenter and self-educated engineer with little formal education, who continued the work done by Hippolythe Pixii, Pacinotti, and others. In 1856 he went to Paris. It was in Froment's workshop, where he was artisan, that Zenobe Gramme met the Italian physicist Pacinotti who had come to Paris to buy instruments and was trying to develop his electric motor (C. Blondel, 1997, p. 169). Gramme applied the ideas for self-generation (of the electromagnetic field) developed by Varley, Wheatstone, and Siemens and created a stable and highly efficient current generator with a continuously wound ring armature: the so-called "Gramme ring." His machine had an immediate and widespread commercial success (Beauchamp, 1997, p. 136).

> *The first Gramme machine was exhibited before the Paris Academy des Sciences in July 1871. It formed the subject of several patents in this country, the first of which was №. 917 of 1870, and there were others taken out in the name of one of Gramme's Paris associates, Hippolyte Fontaine. Within a short time, the Gramme organization was manufacturing dynamos in large numbers in Paris, and similar machines were constructed under license in a number of countries, including England* (MacKechnie Jarvis, 1955a, p. 570).

So Gramme filed for a French patent on November 22, 1869, for a period of seventeen years. Gramme also filed for the British patent №. 1.688 on June 9, 1870, and on October 17, 1871, for the US patent №. 12.057. On April 26, 1872, he filed for British Patent №. 1.254 for a magneto-electric generator, and he filed for an Austrian patent that caused him problems later on.

The US patent №. 120.057 became the subject of litigation in August 1883 in the case of *Gramme Electrical Co. v. Arnoux & Hochhausen Electric Co.* This case addressed the subject that a US patent is no longer valid when another, previously issued, foreign patent for the same invention has expired earlier.[60] The "foreign patent" in this case was a "secret" Austrian patent that had been filed on August 17, 1870, and that was issued on December 13, 1870, "for the duration of one year" (to be extended yearly). The Austrian patent was extended nine times, and it finally expired on December 30, 1880—that is, earlier than the expiration date of the "057-patent": November 22, 1886. This patent was supposedly secret because there was no reference to it in the US patent and due to

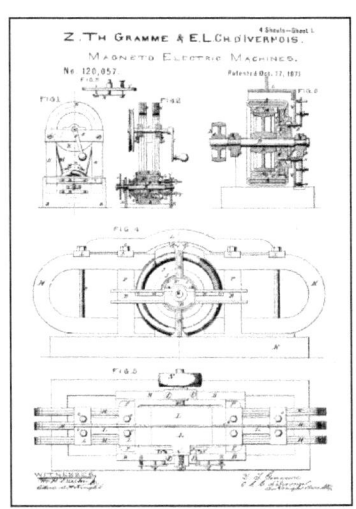

Figure 84: US Patent 120.057 for Gramme's machine (1871)
Source: USPTO

a special clause in Austrian patent law. The Austrian statute provides that the petition for a patent must contain a statement asserting whether or not

---

[60] The same was the case with Edison's patent for the incandescent lamp.

the invention is to be kept secret, and that special care is to be taken for the observation of the required secrecy.

Gramme lost the suit. In his verdict the judge considered the Austrian patent, having expired earlier than the US patent, caused the earlier expiration of the "057-patent" and that there was no case for infringement ("Gramme Electrical Co. V. Arnoux & Hochhausen Electric Co. ," 1883).

Gramme was, next to the electric generator, also occupied with the electric motor—its complete system of generating, distributing, and consuming electricity (the so-called one-to-one systems). In his specifications for US patent №. 269.281 filed on October 9, 1881, it says:

Figure 85: Gramme's dynamos for electroplating (left) and arc lighting (right) (1871)
Source: (King, 1962) pp. 380, 382

This invention has reference more particularly to a system for transferring motive power from one place to another by means of electricity. A system of this kind necessarily involves a generator or generators of electricity for converting mechanical into electrical energy, a conductor or conductors for conveying the electricity, and one or more translating devices or motors for reconverting the electrical into mechanical energy
(USPTO, US patent №. 269.281).

Gramme was looking for finances to fund his patent. In 1871 he founded the company *La Société des Machines magnéto-électriques Gramme* with the help of count Eardley Louis Charles d'Ivernois, and under the management of Hippolyte Fontaine. This became one of the first companies in Paris to manufacture industrial dynamos. Gramme also developed—more or less by accident—a reverse version of the dynamo, the Gramme electrical motor, that became widely used in industry (Beauchamp, 1997, p. 136).

Gramme manufactured in 1872 two different models; for the electroplating applications and for the arc-light applications. One dynamo with low resistance with coarse wire on the armature for electrochemical purposes, and one dynamo of high resistance with fine wire on the armature for use with arc lights. In 1874 he introduced a dynamo "type d'atelier"[61] *to cut down the size and considerably*

---

[61] The dynamo, converting mechanical rotation power into electrical power, always needed a "prime mover." This was often a steam engine, but waterwheels were also applied. So the

increased the efficiency of both the high-resistance and low-resistance generators by relying completely on the principle of self-excitation (King, 1962, pp. 380-384).

By his later improvements, Gramme had converted the electric generator from a laboratory curiosity or an awkward magneto-electric machine into a fully practical dynamo, ready for commercial exploitation. In 1874, four Gramme generators were sold; by 1875, 12 had been sold; by 1876, 85; by 1877, 350; by the middle of 1878, 500; and by 1879, over 1,000. Mechanically, the Gramme dynamo was efficient, compact, and durable; electrically, unlike previous dynamos, it produced a relatively constant output that was greater than that of any previous one, except possibly the Siemens machine. Although the efficiency seems to have ranged between 80 and 90 percent and the main application, until the end of the 1870s, was in the electrochemical industries, the electric light and even the transmission of power was now a possibility (King, 1962, p. 385).

Figure 86: Gramme's industrial dynamo as shown in the Exposition Universelle in Paris (1899)
Source: Internet

Table 8: Some of the British patents granted to Zenobe Gramme

Patent №	Filed	Description
GB 1668	June 9, 1870	Magneto-electric machines/Magneto and dynamo-electric generators (1)
US 120.057	October 17, 1871	Improvement in magneto-electric machines: continuous and alternate induction currents (1)
GB 1.254	April 26, 1872	Magneto-electric machines/Magneto and dynamo-electric generators (1)
US 218.520	August 12, 1879	Improvement in magneto-electric machines: continuous and alternate induction currents
US 269.281	December 219, 1882	Transmission of power by electricity

Source: (Dredge, 1882), Appendix A: Abstracts of patents. Center for Research Libraries, http://dds.crl.edu/ loadStream.asp?iid=17444&f=8; http://dds.crl.edu/loadStream.asp?iid=17444&f=9 (Accessed December 2014)
(1) Patentees Z. Th. Gramme/ E. L. C. d'lvernois

---

combination of a steam engine and an electric dynamo—the so-called "steam generator" or "steam alternator"—was created by generators attached to the flywheels of reciprocating engines.

The Gramme dynamo was successfully introduced in England and in the United States. The Whielden & Cooke Company constructed it under license in England. In the United States, it was introduced as a result of the curiosity of university science professors. The first Gramme dynamo in the United States was built in upstate New York in 1874 under the direction of Professor William Anthoney of Cornell University. Professor George Barker of the University of Pennsylvania acquired a Gramme machine directly from Paris and used it for his lectures calling it the "most perfect dynamo yet invented." Several other machines were acquired by universities and used for educational purposes (J. E. Brittain, 1974, p. 109). The American John Hopkinson and his younger brother, Edward, also undertook a redesign of the Gramme dynamo during 1883, a mutation which became known as the "Manchester dynamo" (Figure 87).

Figure 87: Gramme's Manchester dynamo
Source: Universite de Toulouse, Collection ENSEEIHT

Gramme presented his dynamos at the International Exposition of Electricity[62] of 1881 (the first exhibition totally devoted to electricity) and at the Word Exhibition of 1889 in Paris.

> *Following the Gramme machines, we have a development of alternators by de Meritens, Hefner-Alteneck, Ganz, Schuckert, Zipernowski & Deri, and, in more recent years, such machines as designed by Kapp, Mordey, and Ferrenti* (Rushmore, 1905, p. 254).

Charles Brush's dynamo (1877)

The American Charles Brush (1849–1929) had a great interest in science, particularly with Humphry Davy's experiments with the arc light. He tinkered with and built simple electrical devices such as a static electricity machine at age twelve, experimenting in a workshop on his parents' farm. Brush attended Central High School in Cleveland where he built his first arc light, and he graduated from there with honors in 1867. After graduating from the University of Michigan with a degree in mining engineering, he started his career as an analytical chemist/consultant. In 1876 he secured the backing of his friend George Stockley at the Telegraph Supply

---

[62] From August to November 1881, the first International Electrical Exhibition was held in Paris at the Palais de l'Industrie with an associated international congress. Great advances in electrical technology, especially in electric lighting, had been made, and the exhibition was the showcase for a new branch of engineering: electrical engineering.

Company in Cleveland to design his "dynamo" for powering arc lights. Brush began with Zénobe Gramme's dynamo design, but his final design was a marked divergence, retaining the ring-armature idea that originated with Antonio Pacinotti. Brush obtained a patent for his design on April 24, 1877, US Patent №. 189,997 (Figure 88).

Getting a patent was one thing; maintaining the rights was a totally different story. But there were more difficulties, and those related to the British patent situation:

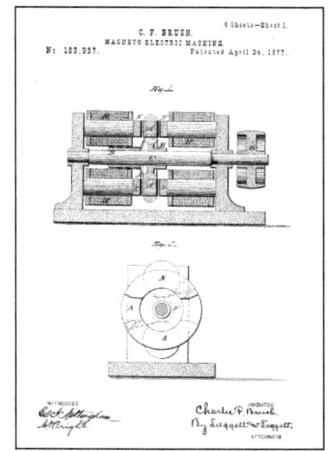

Figure 88: US Patent No. 189.997 for Brush's electric dynamo (1877)
Source: USPTO

> *Notoriously, a UK patent was only as good as its first judicial endorsement, and conditions were ripe in the 1880s for litigation in the electrical field to rise exponentially. The boom years of the British electrical industry coincided with the general surge in patenting consequent upon the Patents, Designs, and Trademarks Act of 1883, which reduced the initial cost of filing a patent from £25 to £4…*
>
> *Increasingly, therefore, in this free-for-all the enforcement of a patent might necessitate recourse to the law…[It was] a highly competitive sector dominated…by large manufacturing firms that were subsidiaries of major foreign corporations…such as Siemens, British Westinghouse, and British Thomson Houston. Their strategy was based on inventions transferred into Britain and reinforced by highly monopolistic patenting regimes, which they could afford to manage aggressively through litigation (or simply the threat of it)* (MacLeod, 2012, pp. 333-334).

In other words, the Americans thought they could outdo Europe by applying their business strategies and tactics. This is exactly what Brush did in England in 1878, as he also patented his invention titled: "Improvements in apparatus for the generation and application of electricity for lighting, plating, and other purposes," as reported in the *London Gazette* on June 7, 1878. The patent was granted with the name of Brush's UK patent agent, Herbert John Haddan. The rights were then transferred to the Anglo-American Brush Electric Light Corporation Ltd. When he filed for an amendment in 1885, both the Siemens and Crompton companies challenged him. But the case was judged in favor of the Anglo-American Brush Company. And in 1886 the Brush corporation started legal action against infringers. Crompton Co. had to agree to pay royalties and take licenses. But Brush's victory did not last long. In October 1888 the Scottish company King, Brown and Co., filed a case against the Brush Co. to prevent any extension of the Haddan-Brush patent. It involved S. A. Varley, who claimed to be the original inventor of the self-exciting and compound-winding generators, as a key witness.

Table 9: Some of the patents granted to Charles F. Brush

Patent №	Granted	Description
US 189.997	April 24, 1877	Improvement in magneto-electric machines: improvement in the armature and arrangement of commutators
US 212.183	February 11, 1879	Electric light-regulator: adjustment of upper carbon rod by the force of gravity and the lower carbon regulated by an electromagnet
US 224.511	February 17, 1880	Dynamo-electric apparatus: dynamo combined with a device for primarily varying the strength of the current
US 260.652	July 4, 1882	Dynamo-electric machine: suppression of extra current in the helices of the field of force magnets
US 285.457	September 25, 1883	Armature for dynamo-electric machines: construction of plates for armature rings
US 312.807	February 24, 1885	Armature for dynamo-electric machines: improvement of patent 285.457
US 376.630	January 17, 1888	Electromagnetic device: protection against large current when carbon rods in arc lights are in contact
US 428.742	May 27, 1890	Commutator collector or block for dynamo-electric machines: commutator blocks
US 746.452	December 8, 1903	Method of manufacturing gas

Source: USPTO

> *As a self-fashioned "morally principled" inventor, he claimed this credit not for any financial profit but for justice, honour, scientific credit, and prestige. Varley attacked not only the Brush company, but also the contemporary scientific authorities who in their role as expert witnesses for Brush had marginalized him in their reconstructions and testimonies…Furthermore, Varley provided a patriotic motive: "I also feel as an Englishmen that there is a certain amount of disgrace in allowing Americans in this country to take precedence and claim royalties for an invention neither originating with nor perfected by them"*
> (Arapostathis, 2013, pp. 149, 158).

On June 26, 1889, the case was ruled in favor of King, Brown and Co.

Brush lost his following appeals on July 16, 1890, and in December 1891. His patent was declared invalid; "justice" was done, for the British had prevailed over the business tactics of the Americans. But there was more to come. The judge Lord Trayner declared "there is no doubt in my mind of his [Varley's] honesty." And Varley was now established in court as the inventor of compound winding and as a trustworthy witness. But the destruction of the Brush patent did not bring Varley any income. Nobody could henceforth accrue royalties on patents for dynamo winding. The invention was now in the public domain (Arapostathis, 2013).

Some of the later dynamos

As a result of the success of the electric lamps, both the arc light and the incandescent light, the need for electric power increased. So other scientists and engineers in Europe and the United States also started developing dynamos. Here are a few of those inventors.

Thomas Edison's dynamo (1879)

In 1879 the American Thomas Edison, who designed complete power systems for his light project with incandescent lamps, created a dynamo. A key feature of the Edison dynamo was its large bipolar magnets, which gave the generator its nickname, the "long-legged Mary-Ann" (a somewhat rude joke among the all-male laboratory staff). Another of this design was "Long tall Sally" (Figure 89). Edison obtained several patents for magneto-electrical machines in the period from 1878 to 1883—for example US patent №. 219.393 of September 9, 1879, for an "improvement in dynamo-electric machines."

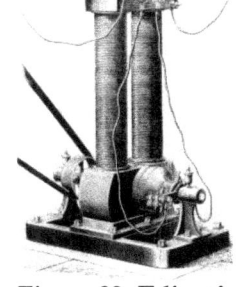

Figure 89: Edison's dynamo "Long tall Sally" (1879)

Source: Smithsonian, (© 2008 IEEE)

Figure 90: Edison-Hopkinson dynamo (1883)

Source: Unknown

It was the consultant John Hopkinson, working on dynamo design, who experimented with models having different types of electromagnet structures. He concluded that the substitution of shorter magnets with increased cross-sectional area would be beneficial to the Edison bipolar machines. In 1883 a dynamo often called the Edison-Hopkinson dynamo implemented the changes he recommended and could supply about twice the number of lights as an Edison dynamo of the same overall weight (Figure 90).

Emile Bürgin's dynamo (1884)

The Bürgin dynamo originated with Emile Bürgin of Basle, Switzerland, and consisted of a series of four or more squares of iron wire, each carrying four coils (one per side) wound in the conventional ring manner. The machine, which was the subject of several English patents from 1875 onwards, was a great improvement on the early Gramme dynamos, in that the "rings" were spaced to produce a machine greater longitudinally than the Gramme—the coils being relatively small in diameter and therefore less subject to heating in operation (MacKechnie Jarvis, 1955a, p. 570).

Willem Smit's dynamo (1885)

On September 22, 1885, the Dutchman Willem Benjamin Smit, working together with Adraan Pot, was granted the US-patent № 326.796 for his Dynamo Electric Machine. Smit was a Dutch entrepreneur working without licenses from other inventors. He copied them (for example Gramme's machine) and also created his own machine. He developed a combined machine (steam unit and electric unit) that could be used as a power station (Figure 91). He also created a dynamo/steam combination and made it into a transportable version (Figure 92). In this way he could create electricity anywhere and demonstrate his dynamo.

Figure 91: Smits's steam dynamo for ship lighting (1900)

In 1883 he started installing electrical incandescent-light systems with Swan lamps on boats traveling the rivers of Holland. He also installed lighting systems in the Dutch colonies—like in 1895 when he installed the electrical power station in Tandjong Priok in the harbor of Batavia (Indonesia). His company *Electrisch-Licht-Machinen Fabriek Willem Smit & Co*, established in 1882, was merged with other Dutch companies (Hazemeyer, Heemaf, Coq, and EMF) into Holec NV in 1969.[63]

Figure 92: Smits's transportable dynamo (1883)

Source: www.willem smithistorie.nl

---

[63] The author of this case study worked from 1976–1981 at the headquarters of Holec. For more information about this Dutch company, see: http://www.willemsmithistorie.nl.

Sebastian Ferranti's dynamo (1883)

The Englishman Sebastian Ziani de Ferranti (1864–1930), an electrical engineer, started working for Siemens (UK) in June 1881 at the age of seventeen. It was there that he met Alexander Siemens, Werner Siemens's brother. He created an AC-dynamo in 1883, after leaving Siemens (Figure 93). The device was noted for its compactness and for its capacity to produce five times more power than any other machine of its size. His unipolar dynamo-electric machine was patented in England (UK patent № 5.926 December 29, 1883) and in the United States (US patent №. 341.079, May 4, 1886). Between 1882 and 1927, he took out 176 patents for a range of products including alternators, circuit breakers, transformers, and turbines. He created a company, Ferranti Limited, that grew into a giant enterprise with more than forty factories and offices across Britain, employing twelve thousand people.

Figure 93: Ferranti's dynamo (1883)
Source: Manchester Museum

From Component to System

The early electrical engineers realized rapid improvements. Some of these people were Frederick Hale Holmes, who began to develop the steam-engine-driven magneto and to use it to illuminate a lighthouse, taking out three UK patents between 1856 and 1857 (№'s 573, 1998, and 2628); Gramme, who created successful dynamos for use in industrial applications (i.e., Gramme's dynamo *type d'atelier*) (Figure 94); and Henry Wilde (1833–1919) who created machines generating enormous electric currents (Wilde, 1867).

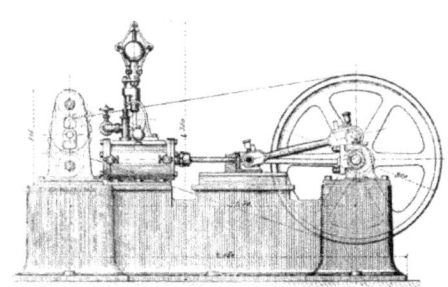

Figure 94: Gramme's dynamo *type d'atelier* combined with a steam engine (1877)
Source: (King, 1962, p. 386)

> *Thereafter, the primitive dynamo was progressively improved by Siemens, Gramme, Edison, Hopkinson, and others until it became an efficient and marketable commodity* (Cardwell, 1992, p. 487).

Over time, thanks to the contributions of many engineering scientists, the dynamo improved in performance, and the builders better understood how to design and make them. These contributions were made not only by the inventor-entrepreneurs such as Edison, Gramme, and Siemens but also by scientist-engineers and by the Brit John Hopkinson (1849–1898), professor in electrical engineering at King's College, London, author of a classic paper *Dynamo-electric machinery* (Hopkinson & Hopkinson, 1886).

Figure 95: Edison's Jumbo Dynamo No. 1

Source: Department of the Interior, National Park Service (1881) www.nps.gov/archive/edis/edisonia/graphics/15400008.jpg.

From electricity generation to electricity distribution

The development of the "dry battery" (the electric dynamo) created an alternative to the "wet battery" (the electrochemical battery based on Volta's discovery). It was a breakthrough of the barrier created by electrochemical batteries.

Soon, electrical generators were applied on an increasing scale, delivering electrical power for individual or local use in totally new applications (Figure 94). Electricity became available (at a cost) in abundance. It was only limited by the "prime mover": the steam machine, the waterwheel, or (later) the water turbine. It rapidly expanded with many applications for electrical illumination that replaced the gas-based lighting systems of those days. Electric lighting started with specific applications in powering military searchlights and powering the strong electric lamps in lighthouses. Next came the big dynamos that powered electric streetlights and the electric lighting of restaurants, theaters, train stations, and big shopping malls—all powered by increasingly powerful dynamos (Figure 95). The development of the dynamo was like a two-sided coin: on one side there was the enormous potential in application, and on the other side, there was the increasing availability of electricity created by continuously improving electric dynamos. But that development also brought a new requirement. Expansion was not only dependent on the development of the component (such as the dynamo and the electric lamp). It became increasingly dependent on the development of the system.

In terms of electricity generation, over a period of about a century the focus of attention had changed from "static electricity" (including the Leyden jar) and "voltaic electricity" (electrochemical batteries) into the creation of electricity by dynamo-electric machines—

Figure 96: Electric dynamos at the International Exposition of Electricity (1881)

Source: http://www.gr-univers.fr/univers/12.php

phenomenon widely presented at exhibitions (Figure 96). And, once introduced, dynamo-generated electricity rapidly took over. Electricity found its application in the form of local systems everywhere, as will be described in the next chapters. For example a factory could have a steam engine that powered the electric dynamo, supplying electricity for arc lights, electroplating, or for powering a central electric motor to power the shaft- and belt-system.

Once the "generating of electricity" problem was more or less solved, the problem became the *distribution* of electricity. Electricity had to be distributed through an infrastructure of electric copper cables. At first the electricity was distributed through 'local' distribution networks, in which a locally erected electric dynamo would power nearby electric lights and motors. It did not directly lead to a revolution, as the application of motors in the time of direct current systems proved to be slow. This was because of the nature of this type of motor, and it was also due to the lack of electricity being available on a larger scale. Before the electric motor could be implemented en masse, another barrier had to be overcome: how to distribute electricity over larger distances. The goal was to find a way so that the "point of generation" of the electricity (close to the primary mover, like water sources) could be a greater distance from the "point of consumption" (the factory or town where the electricity would be used).

The invention of the magneto-electric dynamo

The creation of the magneto-electric dynamo was certainly remarkable and would prove to be fundamental to the further development of electricity in our society. As illustrated, many specific developments took place over time, and each added more to the improvement of all the different machines generating electricity. All these developments had their own relative importance, but there is one remarkable invention that stands out because it definitely had an impact: the conception of the "self-exciting" dynamo.

Looking from a *legal point of view* at the patent situation, it is clear that the patents that were issued for early self-exciting inventions do not give much of an indication that there was a single inventor. It was a range of individual efforts that led up to the self-exciting dynamo.

> *Although the principle of the dynamo was clearly embodied in the Hjorth patent [GB 12.295 of October 26, 1848], its value was not appreciated until sometime later. Eleven years later Wilde employed a small machine with permanent magnets to excite the coil-wound field magnets of a larger machine. But Siemens (British Pat. №. 261 of 1867), taking up the principle employed by Hjorth, dispensed with his superfluous permanent magnets, having found that the residual magnetism, which always remained in iron which has once been magnetized, was sufficient as a basis to start the building-up process. Farmer, Wheatstone, and Varley also recognized this fact about the same time. Siemens's patent also was the first embodiment of what is known as the bobbin armature. Gramme and D'Ivernois (British Pat. 1,668 of 1870, and US Pat. №. 120,057, of Oct. 17, 1871), were the first to bring out the continuously wound ring armature* (Byrn, 1900).

Concerning the controversy around the Albert medal, Wilde might have been the first to publish, build a prototype (of a specific version—with the second motor—of the dynamo), and patent it both in England and in the United States. It showed the principle of *accumulation by successive action*, by combining two of these cylindrical armature machines. His contribution can be considered to be an important one. But he was certainly not *the* inventor of the dynamo-electric machine, when we look at the work done and the patents obtained by others—among those Sören Hjorth and the work resulting in Varley's, Siemens's, and Wheatstone's self-exciting dynamos.

From a *legal point of view*, the development of the self-exciting dynamo-electric machine was an important and contentious one—as can be concluded already from all the discussion (i.e., in legal debates during litigation and infringement cases) about who invented the self-exciting dynamo. The objective of the debates was not so much in financial terms (to obtain a legal and patentable position), but it was more about recognition for being the intellectual owner of the invention.

Looking from a *technical point of view*, the development of the self-exciting dynamo was quite fundamental. And different people at different places accomplished it.

> *The step from magneto-electric to dynamo-electric machines was due to Dr. Werner Siemens, Sir Charles Wheatstone, and Mr. S. Alfred Varley, who quite independently discovered and worked upon the same principle of accumulation by mutual action, the priority falling to Dr. Siemens by previous publication…This principle of accumulation by mutual action is now employed in all machines where currents of great intensity are required* (Higgs & Brittle, 1878, p. 38).

From the *impact point of view* the self-exciting dynamo was enormous. Now electricity became available in an abundance never seen before. It would power the exploding electrical illumination—feeding the electrical motors that were going to be used in a wealth of domestic and industrial applications. Markets for electrical applications exploded due to the availability of electrical power. However, it was not the early scientists and "electricians" who created the business that changed the field of electricity. It was the later industrialist, inventor-entrepreneurs, such as Werner von Siemens (Germany), Smit (the Netherlands), Ferranti (England), Gramme (France), and Brush and Edison (United States), and their lesser known fellow entrepreneurs, who realized the importance of the marketable, affordable, and reliable self-exciting magneto-electric dynamo.

One thing can be concluded from the stepwise invention of the self-exciting electric dynamo; it was a technology that evolved from the minds and hands of many inventive contributors. Having said that, it seems that the technological efforts of Varley, Wheatstone, and Siemens were the most vital to the development of the dynamo.

A cluster of innovation for the dynamo

Compared to the cumbersome wet cell (the nickname for the electrochemical battery), the early dry cell (the nickname for the electric dynamo) was certainly an improvement. It was the result of experimenting with the electromotive principle and the motor-dynamo reciprocity: when feeding an electric motor with electricity, it would rotate, but rotating the same device would result in the generation of electricity. This resulted in the early DC generators.

The breakthrough came through applying principles of the self-eliciting dynamo. It took a while, but the combined—although independent—efforts of Siemens in Germany, and Varley and Wheatstone in England, resulted in a new type of electricity generator: the *self-exciting dynamo* (Figure 97). This dynamo would become the workhorse of the electricity generation. And as a result of that electromotive engine, electricity would become widely available—although the process of bringing electricity into private homes and offices took some decades to be realized.

Over time the electromotive engine of the self-exciting dynamo would be perfected and brought to market by numerous inventor-entrepreneurs (such as Gramme, Brush, Siemens, and others). It would create a range of industries—from the "network" companies distributing electricity to the manufacturers of equipment. Nearly every country would have its own electrical industries that covered its home market. And all these inventors tried to protect their interests with patents.

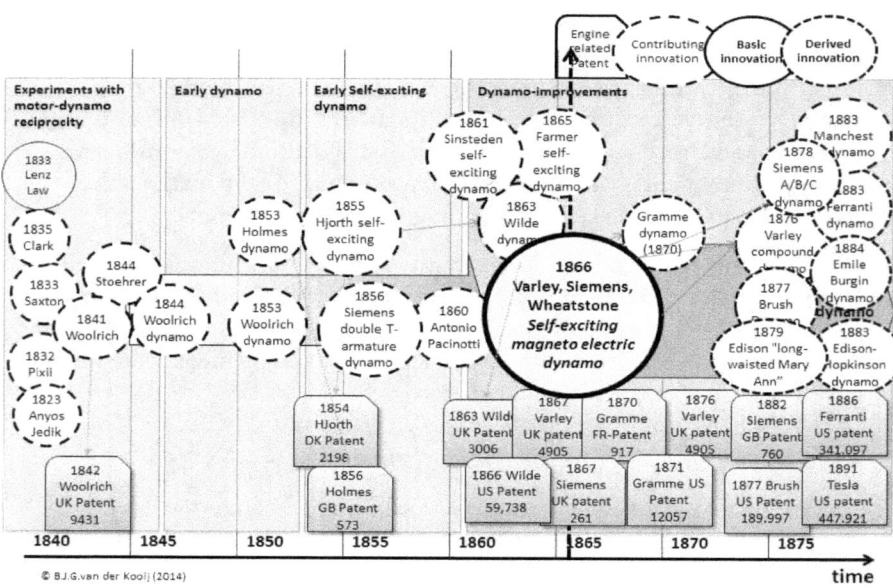

Figure 97: Cluster of Innovation around the Varley, Siemens, and Wheatstone self-exiting dynamo

Source: Figure created by author

Patent activity

All the described activities, experiments, and developments of the magneto-electric dynamo have resulted in a range of patents indicating the extent of the innovative activity. In Table 10 an (indicative) overview is given of those patents that can be considered as important to the early development of the magneto-electric dynamo up to the self-exciting dynamo in 1866. Early patents of a specific inventor are shown, and later patents for the same inventor and foreign patents are indicated.

In Table 11 an overview is given for some of the identifiable patents that were granted *after* the developments around 1866 (those that resulted in the self-exciting dynamo). This table shows the patents issued in the United

Table 10: Overview of magneto-electric dynamo patents 1841–1867

Patent №.	Year	Patentee	Description
GB 9.431	August 1, 1841	John Woolrich	Coating metals/Magneto-electric generator
GB 12.295	October 26, 1848	S.Hjorth	Applications of electromagnetism as motive power/ Magneto-electric generator
GB 2.198	October 14, 1854	S.Hjorth	Magneto-electric battery/ Dynamo-electric generator
D 2.198	October 1854	S.Hjorth	Magneto-electric machine (Danish patent)
GB 573	March 7, 1856	Fr. H.Holmes	Magneto-electric machine /Magneto-electric generator
GB 1.998	July 20, 1857	Fr.H.Holmes	Electromagnetic engines/ Magneto-electric generator
GB 512	February 25, 1859	C.W. Siemens W.Siemens	Electric telegraph/Magneto-electric generator
GB 858	April 1861	H.Wilde	Electro-magnetic telegraph/ Magneto-electric generator
GB 1.994	July–Sept. 1861	H.Wilde	Electro-magnetic telegraph/ Magneto-electric generator
GB 3.006	December 1, 1863	H.Wilde	Electric telegraphs/ Magneto-electric generator
US 59.738	November 13, 1866	H.Wilde	Improvement in magneto-electric machines: current of any desired power
GB 3.394	December 24, 1866	S.A. Varley O. Varley	Generating electricity/Dynamo-electric generator
GB 261	January 31, 1867	C.W.Siemens W.Siemens	Producing electric light at sea/Dynamo-electric generators
GB 1.253	April 17, 1867	C.W.Siemens W.Siemens	Electrical signalling apparatus/ Dynamo-electric generators

Sources: USPTO, (Dredge, 1882); Appendix A: Abstracts of patents. Center for Research Libraries http://dds.crl.edu/loadStream.asp?iid=17444&f=8 (Accessed November 2014)

States and Britain for a multitude of inventions under the label "improvement in electromagnetic engines" and "dynamo-electric generators," respectively.

Table 11: Overview of electric-dynamo patents in 1866–1879

Patent №	Year	Patentee	Description
US 58.960	October 16, 1866	A. P. Berlioz	See note 1
US 63.380	April 2, 1867	C. J. B.Gaume	See note 1
US 69.980	October 15, 1867	W. Wickersham	See note 1
US 78.619	June 2, 1868	L. C. Stuart	See note 1
US 80.463	July 28, 1868	A. J. B.Morat	See note 1
US 96.332	November 2, 1869	G. Little	See note 1
US 103.229	May 17, 1870	H. M. Paine	See note 1
GB 1668	June 9, 1870	Z. Th. Gramme	See note 1
US 105.663	July 26, 1870	L. Finger	See note 1
US 118.561	August 29, 1871	J. P. Tirell	See note 1
US 120.057	October 17, 1871	Z. Th. Gramme	See note 1
US 122.944	January 23, 1872	C. V. Gaume	See note 1
US 126.628	May 14, 1872	M. G. Farmer	See note 1
US 127.369	May 28, 1872	W. H. Richardson	See note 1
GB 1.919	June 25, 1872	C. W. Siemens	Dynamo-electric generators
US 129.000	July 16, 1872	J. S. Camacho	See note 1
US 131.377	September 17, 1872	A. Schreiber	See note 1
US 152.772	July 7, 1874	W. S. Sims	See note 1
US 155.062	September 15, 1874	L. Bastet	See note 1
US 155.396	September 29, 1874	R. van Hoevenbergh	See note 1
US 156.942	November 17, 1874	G. M. Phelps	See note 1
US 166.431	August 3, 1875	A. Tittman	See note 1
US 166.527	August 10, 1875	C. A. Hussey	See note 1
US 171.087	December 14, 1875	J. Bishop	See note 1
US 172.309	January 18, 1876	J. H. Guest	
US 173.561	February 15, 1876	W. E. Sawyer	See note 1
US 187.997	April 24, 1876	C. F. Brush	See note 1
GB 4.905	December 19, 1876	S. A. Varley	Dynamo-electric generators
US 193.385	July 24, 1877	A. Shedlock	See note 1
US 217.807	July 22, 1879	J. C. Ludwig	See note 1
US 219.157	September 2, 1879	E. J. Houston E. Thomson	See note 1
US 219.393	September 9, 1879	T. A. Edison	See note 1

(1): Improvement in electromagnetic engines
Source: USPTO, search criteria: class 310/46; 310/267; 310/265; 322; period 18660101–18793112

Patents were granted to a multitude of inventors who each tried to improve on the self-exciting dynamo. Among those inventors were those who also played a role in the related application fields for electricity—like the electric light and telegraphy/telephony.[64] These men included Sawyer, Farmer, Thomson, Brush, and Edison in the United States and Cook and Wheatstone in England.

The period of 1873 to 1881 was certainly a period of growth for the electric generator. Figure 98 shows the increase in British and American patent activity for electrical dynamos and generators. Figure 99 shows British patents for both the generators and motors. These graphs illustrate that the field of "electrical engines," whether the generator or the electric motor, was becoming a topic of interest for many entrepreneurs. This resulted in a steep increase in patenting activities over the period of a decade to protect inventors' positions in that fast-emerging business. It would also result in many litigation cases. (See Brush litigation on page 120).

As the magneto-electric dynamo is the reciprocal version of the dynamo-electric motor, the development of both machines was parallel. The 1870s were also the time of the revival of the electric motor, which had

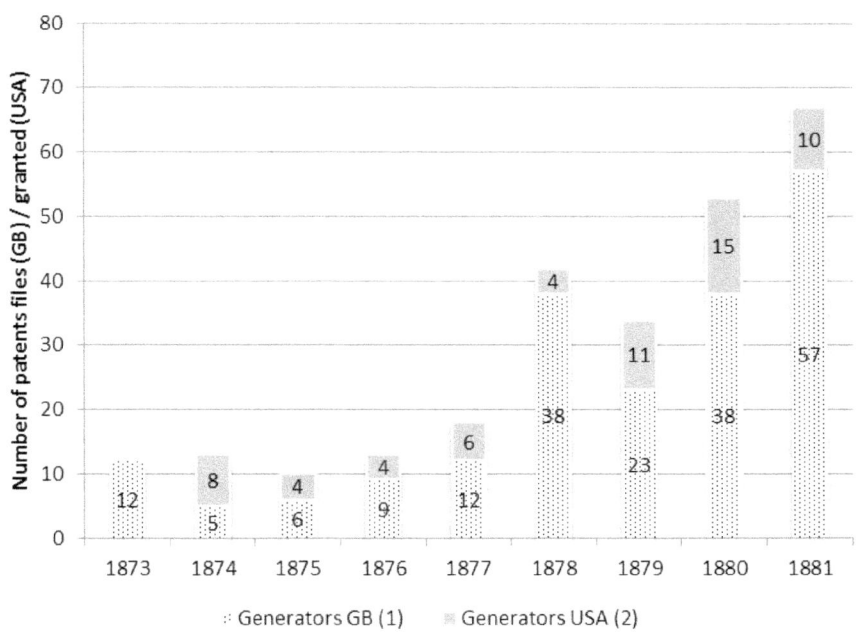

Figure 98: British and American patent activity for electrical dynamos in the period 1873–1881

Based on: 1) (Dredge, 1882). P. Appendix Patents applied for in the United Kingdom, January 1st, 1873- July 1st, 1882. 2) USPTO search: CCL/310/46 or CCL/310/267 or CCL/310/265 or CCL/322.

---

[64] See the separate case studies for specific roles in these applications: B.J.G. van der Kooij, *The Invention of the Electric Light*, *The Invention of the Communicating Engines*. (2015)

been waiting for this new source of electricity. Patent activity for the electric motor followed the patent activity for the electric generator —as can be noticed in the specifications of many of the aforementioned (Great Britain) patents where the same patent could cover both the dynamo and the motor. Figure 99 shows the increase in patenting activity.

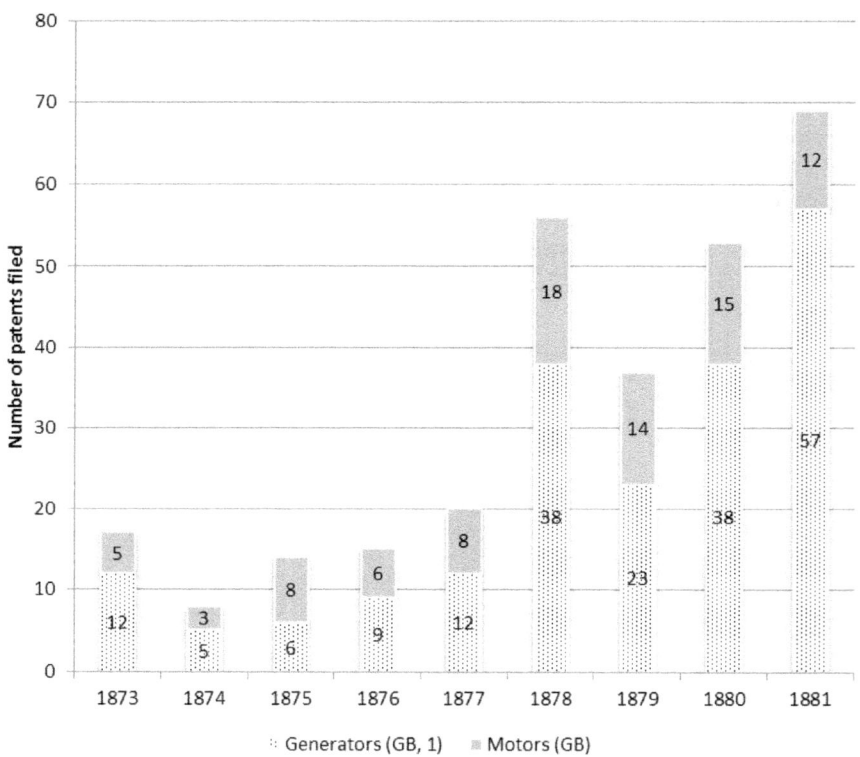

Figure 99: British patents for electric generators and electric motors

Based on: 1) (Dredge, 1882) p. Appendix Patents applied for in the United Kingdom, January 1st, 1873 - July 1st, 1882

The Electric Revolution

In the preceding pages we looked from a technical perspective at the creation and development of the electric dynamo. Now we will turn to a complementary look, the socioeconomic perspective, of the effects of the phenomenon. This perspective includes the business creation and the business context.

Industrial bonanza: cluster of businesses

It was in the decade after the appearance of the self-exciting machines that the development of powerful dynamos started. Companies that were created by the inventors—such as Siemens, Brush, Gramme, et al.—started supplying a range of different dynamos for a range of applications (i.e., electroplating, lighting, and transportation). Others soon followed this activity by taking either a license on a patent or just by trying to copy the more successful machines. It was not only the suppliers of dynamos, and later the arc lamps and the incandescent lamps, who were part of the

Table 12: Some inventors and their first companies*

Inventor	First Company	Activities/Patent
W. Siemens	Telegraphen-Bauanstalt Siemens & Halske (Germany, 1846)	Manufacturing electric dynamos: Dynamo, GB-Patent №. 261 (1867)
Z. Gramme	La Société des Machines magnéto-électriques Gramme (France, 1871)	Manufacturing electric dynamos: Dynamo, GB-patent №. 1.688 (1870)
C. F. Brush	Telegraph Supply Company (United States, 1876)	Dynamo and arc-light development: US-Patent №. 189.997 (1877)
R. E. B. Crompton	Crompton & Company (United Kingdom, 1878)	Manufacturing arc lamps and the Burgin dynamo: installation of big lighting projects
E. Weston	Weston Dynamo Machine Co. (United States, 1879)	Manufacturing electric dynamos, in 1880 renamed Weston Electric Light Company, manufacturer of arc lights
E. Thomson	American Electric Company (United States, 1880)	Manufacturing of arc lamp and dynamos In 1883 Thomson-Houston Electric Company, manufacturer of arc lamps and incandescent lamps
S. E. Ferranti	Ferranti, Thompson and Ince Ltd. (Great Britain, 1882)	Manufacturing DC-dynamo generator and arc lamps
W. Smit & A. Pot	Elektrisch-Licht-Machinen Fabriek Willem Smit & Co (Netherlands, 1882)	Manufacturing dynamos, lamps, and ornaments

* These are just a few examples of the business development that took place around the manufacturing of electric dynamos and related equipment (like switching equipment).

evolving industry, but it was also the suppliers of additional equipment (switches, isolators, circuit breakers, meters for measuring the consumption of electricity, and so forth). This resulted in the industry of electrical-equipment manufacturers that started around the 1870s. The industry saw a lot of fierce competition in its infancy, which would result in many mergers and acquisitions and would lead to a couple of dominant manufacturers.

In addition to these *equipment manufacturers* (that made dynamos, electric lights, and related parts), companies were created for the distribution of electricity: the electricity distribution industry or "utility companies." The distribution networks started to increase in size—even more so when, after about 1887, alternating-current generators came into extensive operation. This, together with the commercial development of the transformer,[65] in time revolutionized the transmission of electric power to long distances. Likewise, the introduction of the rotary converter (in connection with the "step-down" transformer), which converts alternating currents into direct currents, made the merger of AC systems and DC systems possible. All of these developments have largely affected the operation of electric power systems.

Business context: role of finance

In a short period in the second half of the nineteenth century, the electrical manufacturing industry transformed through many mergers and acquisitions. For example, in the United States, there were a diversity of companies in the 1860s but only a couple of large organizations left by around 1900. This was not accidental, but the result of the financial powers that financed the industry. Among those was the merchant banker J. Pierpont Morgan, the "Napoleon of Wall Street":[66]

> *J. P. Morgan became the master of Big Business through control and interlocking boards of directors. These types of financial arrangements became known as "Trusts." In addition to these trust-arrangements, Morgan still functioned as a merchant banker controlling credit. The power and control of Morgan was unbelievable. Even the government, prior to the Federal Reserve Act, depended on*

---

[65] An electric device which "transforms" low-voltage, high-current electricity into high-voltage, low-current electricity, and vice versa.
[66] Among his varied business interests was the International Mercantile Marine, the shipping combine that controlled Britain's White Star Line, owner of the *Titanic*. Morgan attended the ship's launching in 1911 and had a personal suite on board with his own private promenade deck and a bath equipped with specially designed cigar holders. He was reportedly booked on the maiden voyage but instead remained at the French resort of Aix to enjoy his morning massages and sulfur baths. Source: Daugherty, G.: Seven Famous People Who Missed the Titanic. http://www.smithsonianmag.com/history/seven-famous-people-who-missed-the-titanic-101902418/ (Accessed May 2015)

> *Morgan to keep the country fluid. He could also use "financial panics" to take over companies. In those days a bank could call in short-term debt. If the company did not have the cash, the bank could end up owning the stock and bonds of the company…His overriding belief was that competition was wasteful and destructive. He controlled industries to avoid competition and maximize corporate profits. Companies that got in his way could be subdued by drying up their credit in difficult times…no single American businessman ever had the power of Morgan over the economy. He controlled the railroad, steel, oil, electrical,[67] and banking industries, as well as the gold market*
> (Skrabec, 2007, pp. 95,96).

There were other financial factors influencing the development of the electric industry as a whole, including the several financial "panics" that occurred in the second half of the nineteenth century—for example the Panic of 1853, the Panic of 1873, the Panic of 1893, and the Panic of 1907. These crises resulted in a range of bank failures, stock market crashes, businesses that went bankrupt, building construction that halted, and increasing unemployment. The panics of 1873 and 1893 contributed to the Long Depression of 1873 to 1896, a period of worldwide recession. The economic climate also affected the electric industries. It started with the bankruptcy of a bank.

> *On Wall Street, the panic started on September 18, 1873, with the suspension of Jay Cooke and Company. The financier, famous for having marketed more than a billion dollars in US bonds during the war, had invested heavily in railroads, especially a second transcontinental: the northern Pacific Railway. However, in 1873 the [rail]road was nowhere near completion, and Cooke failed to sell new securities in a very tight market. Having underwritten the company, he went bankrupt…Many banks failed in its wake, as credit suddenly withdrew from the market, and short-term loans were recalled. The New York Stock Exchange closed on September 20—for the first time in its history—and did not reopen for the following ten days. At the urgent request of many prominent bankers, the US Treasury injected money into the system, first by buying US bonds, then by reissuing greenbacks. To protect their rapidly depleting reserves, New York banks partially suspended payments on their notes and centralized payments in the New York Clearinghouse (a consortium of banks that issued loan certificates, instead of cash, for interbank transfers). By November, the financial storm had passed,*

---

[67] Morgan, being the largest stockholder of Edison Electric, was also one of Edison's principal financiers. In this context it is not so strange that Edison's Pearl Street project also encompassed Morgan's offices.

and banks resumed payments of their notes...In a month, fifty-five roads had failed to meet their payments, and in three years, half of the railroad companies went to receivership. Railroad construction virtually stopped for the remainder of the decade. This, in turn, dramatically reduced the demand for many industries. Output of iron and steel declined by 45 percent in barely a year. Construction of machines dropped heavily. Production in other sectors was not hit as severely (agricultural output even continued expanding), but economic conditions worsened for them too, especially trade, building construction, and services. New York international commerce dropped. The country experienced the longest contraction of business in its young history (Barreyre, 2011, pp. 406-408).

The silver recession of the late 1880s slowed down economies in the United States and Europe, and this—coupled with skyrocketing increases in the price of copper—sent shock waves through the electrical industry. The high costs of patent litigation and price competition caused the money boys to rethink their strategy. Edison's companies were consolidated and placed under the control of professional managers. The banks called their Westinghouse loans, and it appeared that Westinghouse would go out of business.[68]

Business context: role of government

The economic upheaval just described all happened within the American capitalist context, with a government that was reluctant to intervene. In Britain the economic developments were comparable, but the behavior of the government was different. There, over time, the industry was faced with more and more government regulations. The English government, for example, after experiencing gas-company monopolies making excessive profits, and faced with a bonanza of distribution initiatives, passed legislation in 1882 (Electric Lighting Act) and 1888 (Amended Electric Light Act).

> *In an epoch characterized as "the end of laissez faire" and in a nation whose representative government showed increasing concern for the welfare of the growing body of the electorate, it is not surprising that Parliament, the central bureaucracy, and the local authorities reacted to the intense activity and optimism in the electric-light industry. Within two weeks of the formal opening of Holborn and during the spring speculation, a select committee of the House of Commons considered, and heard testimony on, proposed central station legislation* (Hughes, 1962, p. 30).

---

[68] Source: Metcalf, J.F. The History of Electricity. http//www.electric-history.com/~zero/005-Electricity.htm. (Accessed December 2015)

Legislation certainly influenced the development, but it did not always have the desired effect.

> *After the passage of the Act, after the legislated conditions had been defined, Great Britain—still thought of as the world's greatest industrial nation and distinguished by her scientists and engineers, seemed ready to move ahead with her central station industry. Economic conditions were favorable: 1882 fell within a limited period of recovery during the "great depression in Britain, 1873–1896."…By year's end, 1884, the doldrums had set in. Bad times had not come upon the electrical industry alone. Innovators of the electric light and central station industry found their melancholia echoed by financiers and managers from the older trades of ship-building and heavy metals. Although 1882 had come at the end of the recovery cycle within the "great depression," 1883 and 1884 fell within a slump. Yet spokesmen for the private enterprise sector of the electrical industry were reluctant to view the problems of the industry in the broad perspective of the economy; they preferred to limit analysis to government activity…*
>
> *Since investors had put £7,000,000[69] in electric company shares in 1882 alone (reports of progress in the central station industry in the USA had been comparatively bright) and since influential persons were interested in the electrical industry, Parliament considered amendments in 1886 to the Electric Lighting Act…Other explanations for the lagging central station industry had spokesmen during the time of its doldrums, but after all were heard, private enterprise had made its case most effectively: the Electric Lighting Act of 1882-government legislation, had paralyzed the central station industry* (Hughes, 1962, pp. 31-35).

Over time the supply of electricity to homes, offices, shops, factories, farms, and mines became the responsibility of public utilities, which were either private organizations subject to monopoly regulation or public authorities owned by local, state, or national bodies. It was not before the early decades of the twentieth century that these industries stabilized. In 1926, British Parliament passed the Electricity Supply Act, that resulted in the construction of the National Grid in England from 1929 to 1932.

---

[69] Calculated on the economic power value, that amount would be equivalent to £8,642.000.000 in 2013. Source: http://www.measuringworth.com/ukcompare/relativevalue.php.

The same thing happened all over the world. It was the electrification of the world, driven by enormous demand for electrical applications—not only equipment for the suppliers and distributors of electricity, but also for industries supplying products using electricity. These included, among other things, electric home appliances such as washing machines, refrigerators, and electric flatirons (Figure 100).

Figure 100: A pile of two thousand flatirons exchanged for new electric irons, ca. 1912
Source: General Electric Collection at the Schenectady Museum & Suits-Beuche Planetarium

Booming markets

After all the efforts of a) the experimenting scientists who discovered the nature of electricity, b) the engineering scientists who developed the

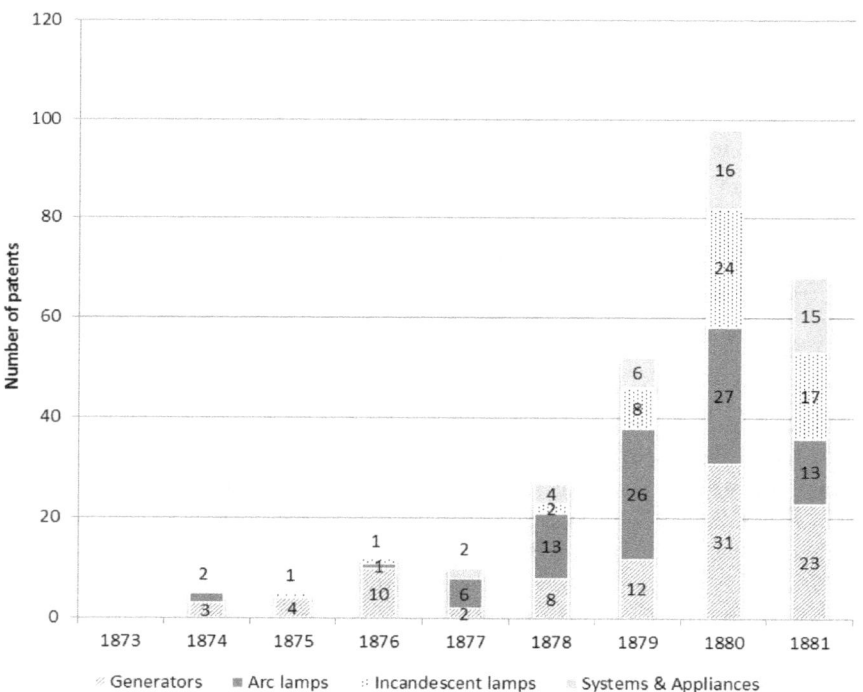

Figure 101: US patents for electric generators, incandescent lamps, arc lamps, and systems and appliances in the period 1873–1881
Source: (Dredge, 1882), Subject Matter Index, Patents related to Electric Lighting

electromotive engine and dynamo, and c) the theoretical scientists who gave insight into the phenomenon of electricity, the world was at the threshold of the "Electric Revolution" around the 1880s. "Ours is the age of electricity," observed the editors of *Electrical World* in 1883, "everywhere electricity is fast becoming the all-inspiring, all-controlling influence. It may be said to be 'fashionable' in the extreme just now as the most popular agent at the disposal of man. It fills everybody with interest and curiosity" (Dalzell, 2010, p. 37).

Indeed, new technologies in telegraphy, in arc lighting, in incandescent lighting, in electric motors, horseless streetcars, telephony, phonographs and motion pictures, power generation and transmission—a host of experiments and applications—were appearing everywhere. These technologies were transforming the material landscape and attracting

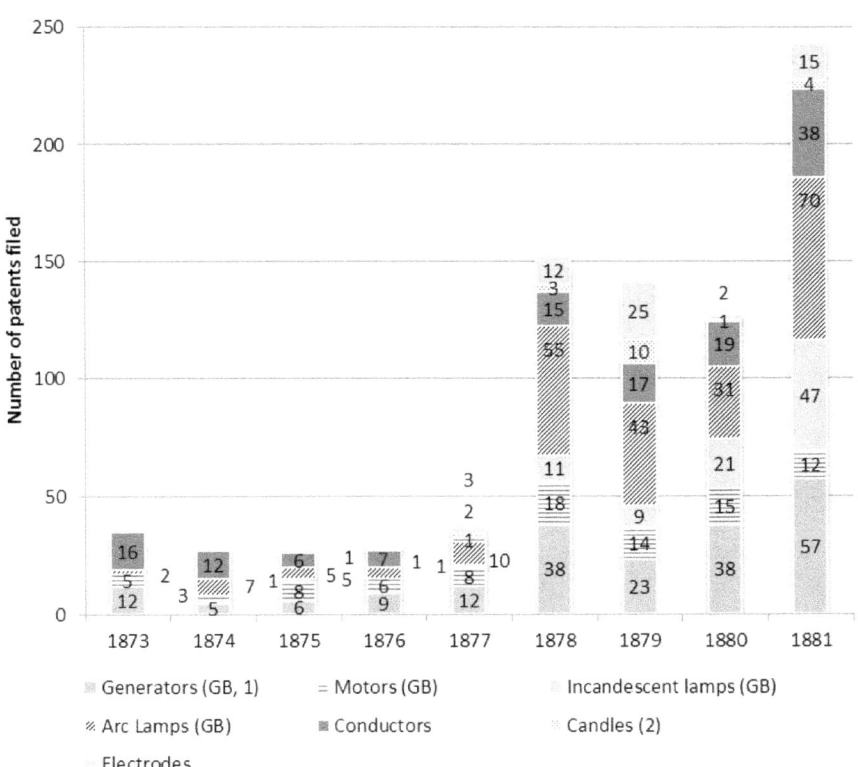

Figure 102: British patents for electric generators, motors, incandescent lamps and arc lamps in the period of 1873 to 1881

Source: (Dredge, 1882), Subject Matter Index, Patents related to Electric Lighting
(1) Including self-exciting generator (2) candle lamps, carbon holders

inventors and entrepreneurs—the electricians and the "wizards" of those days—by the dozens, hundreds, perhaps thousands.

The increasing interest in everything related to this magical, new form of power was reflected in the number of patents that were filed. In the United States, between 1870 and 1895, the US Patent Office issued over 17,500 electrical patents. In the period 1873–1881, and again in an 1882 analysis, hundreds of patents were issued (see Figure 101) (Dredge, 1882). In Britain during 1873–1881, in an 1882 analysis, more than one thousand patents related to "electric lighting" were filed. Many of them were filed for generators, incandescent lamps, and arc lights. But there were also patents issued for supporting parts, such as cables and cabling systems and electrodes. (See Figure 102.)

Many of all those patents were for the electric lamp—both the arc lamp and the incandescent lamp. And of these inventors, it was Edison who was the king of invention with about 1,100 patents. Among those were over four hundred for "electric light and power" (Edison, 2013, p. Edison's Patents).

As will be explained in a subsequent case study,[70] after their original inventions, the arc lights were more or less fully developed by 1876 (Yablochovich, et al.), incandescent lights by 1882 (Edison, et al.), AC-induction motors by 1890 (Dobrovosky, et al.), DC-powered electric streetcars by 1890 (Sprague, et al.)—each within some five years after major innovators began their efforts. As it is impossible to cover the whole field of applications of electricity, we will limit ourselves to the one application that proved to be a mighty impetus to the further development of electromotive power: the application of electric light and electric rotative power.

Early electric systems

With the advent of the generator, electricity passed the barrier of the cumbersome voltaic cells. Electric power became available in ever-bigger quantities. Its supply was only limited by the prime mover: steam power or waterpower. One of the early applications for DC electricity was electroplating—technology in which metal (nickel, chrome, silver, or gold) is dissolved in a solution and deposited on another surface. Although this process had already been applied earlier (with batteries supplying the current), the arrival of the generator gave electroplating a boost. And it showcased the advent of the early electric systems.

---

[70] See: B. J. G Van der Kooij, *The Invention of the Electric Light* (2015).

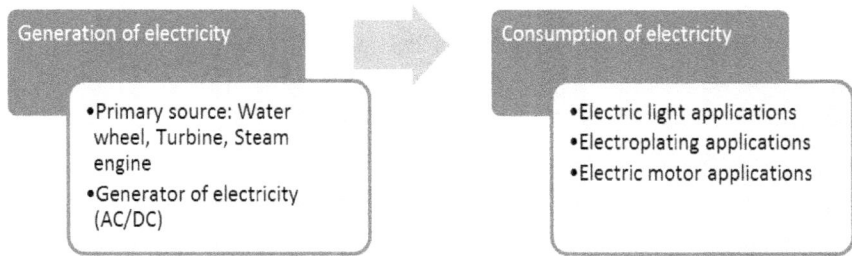

Figure 103: Overview of the early DC-electric systems of generating and consumption of electricity

The basic electric power system (Figure 103) consists of a "generating" part and a "consumption" part.[71] It is the electric generator that supplies the electricity to a "load." In the case of electroplating, the load is the electroplating bath with its anode and cathode. In the case of the electric light and electric motor applications, it is the lamp or the motor that is the "load."

Depending on the application, the configuration of this basic system could be a *one-to-one* system: one generator supplying electricity for one light (for example in a lighthouse). Or it could be a *one-to-many* system (like a generator supplying electricity for a range of arc lamps). Over time, systems became a little more complicated, though.

The era of power

The *Era of Light* [72] had resulted in the broad acceptance of electric lighting systems. Next, electricity was used for some special industrial applications such as electroplating and heating. But that was all; electricity was not yet widely used as a source for rotative applications.

As explained before, the introduction of the DC motor was a slow process. Occasionally DC motors would be developed and patented for specific applications (for example US patent 12.106 granted to Louis Stein for a revolving ceiling fan on December 19, 1854). But in the 1880s, in the same time frame as the developing applications for electric light, some interesting developments took place, which included the use of the electric motor.

---

[71] Generally speaking, electricity cannot be stored. The quantity of electricity generated (current multiplied by voltage) has to be consumed directly. The (limited) storage of electricity in rechargable batteries is the exception.
[72] See: B. J. G. van der Kooij.*The Invention of the Electric Light* (2015).

Scores of small clothing factories and tailor shops in Boston used batteries of sewing machines operated laboriously by hand. Larger establishments were equipped with machine-tables, which had a shaft connecting all the sewing machines, so that they could be run by power. In December 1886, the first 220-volt Sprague motor was installed in a building at Purchase and Pearl Streets, Boston, for the purpose of running a freight elevator. The motor was a fifteen-horsepower unit, connected by about three thousand feet of copper wire to the three-wire system of the Boston Edison Company. At the end of 1886, the Sprague Company had 190 stationary motors installed and in use, and 80 more under construction. In many cities in the East and Middle West, they served more than a hundred trades and industries. They drove boot and shoe machinery in Detroit and Boston; coffee mills in Elgin, Illinois, and Lancaster, Pennsylvania; emery wheels in Des Moines and Chicago; lathes in Chicago, Boston, and New York; and printing presses, ventilators, ice cream freezers, and various other mechanisms. The Chicago fire department; the Gold and Stock Telegraph Company of New York, where Edison started his career as mechanical repair man; Drexel, Morgan and Company; and the New York Stock Exchange installed Sprague motors (Hammond, 1941, pp. 118-119).

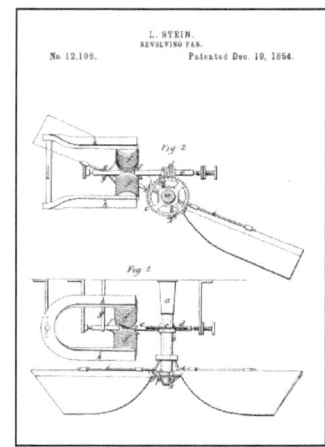

Figure 104: Electric ceiling fan patented by L. Stein (1854)
Source: USPTO

Mobile applications of DC motors

Those were the *stationary* applications of the electric motor. But, just like when the steam machine was introduced, the electric motor was soon going to be used in *mobile* applications—like to create "electric railways." It was Frank J. Sprague (1857–1934) who played an important role when he started his Sprague Electric Railway & Motor Company in 1884.

Frank Sprague (1884)

Born in Milford, Connecticut, Frank Sprague (1857–1934) was raised by his aunt Elvira Sprague in North Adams, Massachusetts, after the death of his mother when he was eight years old. North Adams was becoming quite an industrialized and economically expanding community, and offered the young boy lots of opportunities to find small jobs. "While attending…High School," he later recalled, "I tried to add to the [household's] meager income, selling lemonade from a can carried by a shoulder strap, or apples from a basket to shoe shop workers, as well as collecting newspaper and doctor's bills and soliciting orders for papers and book bindings" (Dalzell, 2010, p. 27). After he attended Drury Academy (a local private preparatory school), he was admitted as a cadet to the US Naval Academy at Annapolis, Maryland in 1874 and graduated four years later as "passed midshipman."

> *The formal academic training in both physics and mathematics provided him with a sound theoretical approach to grasping electrical technologies. At the same time, the practical, problem-solving framework prepared him for the concrete, mechanical challenges of invention, including assembly, improvisation, and refinement of designs. Sprague came out of Annapolis equipped with both a fundamental grasp of scientific electrical theory (circa 1878) and a resourceful capability for "craft knowledge" (in the sense of hands-on trial and error) as a means of working toward technical solutions* (Dalzell, 2010, p. 31).

With a group of his classmates, he visited the Philadelphia Centennial Exposition in 1876. There his attention was drawn to the Machinery Hall and the electrical exhibits, where he admired the "state of art" inventions in telegraphy, telephony, and electrical equipment such as dynamos.

> *The electric dynamos exhibited by both the Gramme Electric Company and Farmer-Wallace that supplied the power for arc lighting that illuminated part of Machinery Hall. The Farmer-Wallace dynamo, designed by Moses Farmer and manufactured by Wallace & Sons, represented state-of-the art technology in that the machines were self-excited…More generally, the Centennial Exposition created a sense of technology that Sprague imbibed deeply and definitively* (Dalzell, 2010, pp. 34-35).

After completing classwork at the Naval Academy, midshipmen undertook a two-year cruise before returning to Annapolis for examination and a final rating. Sprague was assigned to the Asiatic squadron and the *USS Richmond*. Returning in 1880 from a tour to East Asia, he worked on the Newport Torpedo Station under the guidance of Moses Farmer (the facility's electrician). Here Sprague developed a compact

Figure 105: The Sprague motor (1884)

Source: The Shore Line Trolley Museum www.bera.org

dynamo around 1881 (and filed for a patent on October 4, 1881, that was granted as US Patent 304.145 on August 26, 1884).

> *The novel construction impressed Farmer, who supported the young midshipman when Sprague applied to the navy for permission to attend an international electrical exhibition in Paris in 1881. Despite Farmer's endorsement, permission was denied. Sprague seized a second opportunity, however. After arranging an assignment on the USS Lancaster in the Mediterranean squadron, he reached Europe and promptly took a three-month leave—too late to reach Paris but in time to attend another electrical exhibition in London early the next year. He arrived, he later recounted, "with about $20 and the necessity of presenting urgent needs to the US Dispatch Agent."[73] Reaching the Crystal Palace, he secured an appointment on the Exhibition's Jury of Awards as secretary for the panel testing gas engines, dynamos, and electric lights…While in London, Sprague met Edward H. Johnson, one of Edison's business partners and managerial lieutenants who was in England to supervise the exhibition of incandescent lighting entry at the Crystal Palace. Johnson, impressed by Sprague's technical knowledge, recommended him to Edison* (Dalzell, 2010, pp. 49-50).

So Sprague resigned from the navy and started working for Edison in Menlo Park in the construction department. There he got hands-on experience in the field and was able to exploit his mathematical capabilities. He developed an electric motor that could be adapted for use with industrial machinery. He also helped with the installation of Edison's groundbreaking three-wire electrical light systems and made refinements to the inventor's power-distribution system. Sprague's apprenticeship with Edison thus gave him opportunities to exercise both his academic training and his engineering skills. And it was his designs for electric motors that would prompt him to resign from working with Edison.

> *The motors that Sprague assembled during this period (the last few months of 1883 and the first few months of 1884) drew in part on existing motor designs as well as Sprague's dynamo ideas…Sprague devised a motor with reverse wiring in proportions that would equip it to operate at the same speed regardless of the size of the load that it was carrying. This design represented an entirely new approach to motor design, with significant implications for application* (Dalzell, 2010, pp. 56-57).

---

[73] In 2013 that would be about $470 (based on the Consumer Price Index calculation). Source: http://www.measuringworth.com/uscompare/relativevalue.php

Edison's focus at the time was on electric lighting, but Sprague believed that the Edison companies were overlooking the many possibilities of electricity as motive power. So Sprague left Edison's employment and started his first venture: the Sprague Electric Railway and Motor Company (in short SERM) in 1884. His decision meant taking quite a risk: moving from the entourage of the famous Edison and venturing in a start-up that addressed a whole new field of applications.

> *He was twenty-seven years old, had no capital and a few resources, and would be challenging rivals who were already building out their own designs and establishing themselves in a rapidly forming market…He capitalized his venture at $100,000, but he had nothing like that sum at his disposal…Then he turned to Edward H. Johnson, one of the partners who managed Edison's lighting company, for substantial backing…he connected Sprague…to potential financiers such as J. P. Morgan and Henry Villard* (Dalzell, 2010, pp. 68-69).

Sprague was in business and continued from 1884 with his ideas for an electric DC-traction motor with the aim of creating an electric street railway. He was aiming at the enormous market of horse-drawn transportation that had started to boom in the 1850s. This boom had two results: increased mobility for ordinary people and cities that were drowning in manure (Figure 106)—and other hazards like urine, flies, congestion, carcasses, and traffic accidents.

Figure 106: Horse manure in the streets of New York (1850s)

Source: http://www.19thcenturybottlediggers.com/sanitation101.htm

> *By 1853 New York omnibuses carried 120,000 passengers per day. Needless to say, this required a tremendous number of horses, given that a typical omnibus line used eleven horses per vehicle per day. And the need for horses was to spiral even further when omnibuses were placed on tracks, increasing their speeds by 50 percent and doubling the load a horse could pull. Fares dropped again, and passengers clamored for the new service. By 1890 New Yorkers took 297 horsecar rides per capita per year…Experts of the day estimated that each horse produced between fifteen and thirty pounds of manure per day. For New York and Brooklyn, which had a combined horse population of between 150,000 and 175,000 in 1880 (long before the horse population reached its peak), this meant*

that between three and four million pounds of manure were deposited on city streets and in city stables every day. Each horse also produced about a quart of urine daily, which added up to around 40,000 gallons per day for New York and Brooklyn...The clatter of horseshoes and wagon wheels on cobblestone pavement jangled nineteenth-century nerves...

Congestion was another problem. Traffic counts indicate that traffic across the nation more than doubled between 1885 and 1905. Not only was the number of vehicles rising rapidly, but the nature of the vehicles

Figure 108: Philadelphia street congestion: Chestnut Street (ca. 1900)

The street is crowded with horse-driven carriages, an electric streetcar, and people. Source: B. W. Kilburn Company (1897), National Archives

themselves caused tremendous problems. A horse and wagon occupied more street space than a modern truck. Obviously, horse-drawn vehicles traveled at very slow speeds, and horses, especially those pulling heavy loads or hitched in teams, started forward very slowly, a great difficulty in stop-and-go conditions. Streets of the era were not adequate to handle the traffic, and hills caused problems (Morris, 2007, pp. 4-6).

The problems with horse-powered street transportation were everywhere, but especially in the big cities (Figure 108). If a solution could be found to get rid of the horse, urban planners and municipal authorities would certainly pay attention to it. And Frank Sprague was one of the people who thought he had a solution: the streetcar powered by an electric motor (Figure 107).

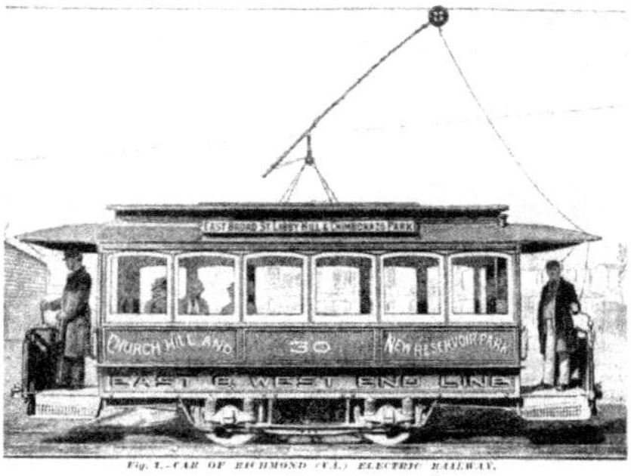

Figure 107: Car of the Richmond Union Passenger Railway (1888)

Source: http://www.cable-car-guy.com/html/ccmanbl3.html

Sprague's commercial success with various electric motors provided him the capital he needed to take on this task. At first he obtains US Patent № 295.454 for a multi-coiled electrodynamic motor with variable speed and the possibility for changing direction of rotation on March 18, 1884. Several other patents would follow this one (for example US Patent №. 313.546 for an electrodynamic motor that was granted on March 10, 1885).

But Sprague did more than just develop a motor; he added the speed/power regulation device. (US Patent №. 313.247 for a regulator for electric motors was granted on March 3, 1885.) He thought outside of the system and devised a supply of electricity that could be accessed through conductors embedded in the street using sections (US-Patent №. 323.459 granted on August 4, 1885). He also designed a new understructure by placing a DC motor on the central axis of a carriage (US-Patent №. 324.892 granted on August 25, 1885). Sprague was clearly designing the whole system for the electric traction of a railway car (Table 13).

Table 13: Sprague patents for generator, DC motor, and railway system

Patent №	Granted	Description
US 295.454	March 18, 1884	Electrodynamic Motor: reversible and variable speed motor for a multi-coiled DC system to be used in series or on movable structures such as railways (filed May 2, 1883)
US 304.145	August 26, 1884	Dynamo-electric machine/generator (filed October 4, 1881)
US 313.247	March 3, 1885	Electrodynamic motor: regulator for multi-coiled motor, improvement on Patent №. 295.454 (filed February 21, 1884)
US 313.546	March 10, 1885	Electrodynamic motor: regulator for multi-coiled motor, improvement on Patent №. 295.454 (filed November 4, 1884)
US 315.179	April 7, 1885	Electrodynamic motor: improvement on Patent №. 295.454 (filed June 9, 1884)
US 317.235	May 5, 1885	Electric railway: conductor system for supplying electricity as a motive power in railways (filed April 14, 1882)
US 323.459	August 4, 1885	Electric Railway System: section based conductor system for supplying electricity as a motive power in railways (filed January 19, 1885)
US 324.892	August 25, 1885	Electric railway motor, combination of wheeled vehicle and electromagnetic motor mounted on driving axle (filed May 25, 1885)
US 328.821	October 20, 1885	Electric railway system: system having two tracks supplying electricity to two parallel motors (filed February 27, 1885)
US 338.313	March 23, 1886	Electric railway system: (filed December 15, 1884)
US 372.822	November 8, 1887	Dynamo-electric machine: winding for the armatures (filed March 3, 1885)
US 397.875	February 12, 1889	Overhead line for electric railways (filed September 27, 1888)

Source: USPTO

By mid-1885 Sprague's company, SERM, was booking sales and delivering motors, and the bulk of their business was used for stationary applications. Continued growth through 1886 resulted in an acceptable cash flow. And by 1886 Sprague's company had introduced two important inventions: a constant-speed, nonsparking motor with fixed brushes and regenerative braking—a braking method that uses the drive motor to return power to the main supply system. His motor was the first to maintain constant speed under a varying load. It was immediately popular and was endorsed by Edison as the only practical electric motor available. His regenerative braking system became important in the development of the electric train and the electric elevator.

It was in late 1887 and early 1888, using his trolley system, that Sprague installed the first successful large electric street railway system, the Richmond Union Passenger Railway in Richmond, Virginia. Sprague had taken a massive risk, as he would be paid ($110.000)[74] only after the system worked successfully and was up and running with thirty cars in operation at a time.

> *This contract comprised "the building of a generating station, erection of overhead lines, and the equipment of 40 cars, each with two 7 ½ horsepower motors on plans largely new and untried." The overhead trolley system under a pressure of 450 volts, with the track rails forming the return circuit, was used*
> (Martin & Coles, 1919, pp. 23-24).

Financially, the project was a loss (estimated later by Sprague at $75,000)[75], but technically, it was his showcase. SERM, however—constantly underfinanced—needed Sprague and his partner Johnson to take personal loans ($45,000 and $40,000 respectively) to keep the company afloat. But then SERM revenues climbed from just under $30,000 in 1887 (from motor sales) to nearly $365,000 in 1888 and $1.5 million in 1889. Exploiting his first-mover advantage, Sprague secured roughly half of the two hundred electric railway projects that went into construction between 1888 and 1890. Sprague had survived the first years of his venture (Dalzell, 2010, p. 95). Later SERM would be part of the restructuring of Edison's companies, and Sprague would lose control over his company.[76] But that

---

[74] This project amount would be equivalent to more than $22 million in 2010, calculated on the basis of labor cost. Source: Measuring Worth at http://www.measuringworth.com/uscompare/ relativevalue.php
[75] This project loss would be equivalent to more than $9 million in 2010, calculated on the basis of labor cost. Source: Measuring Worth at http://www.measuringworth.com/uscompare/relativevalue.php
[76] In 1890 Sprague sold (his share) in his company to the Edison General Electric Company for $750,000 (equivalent to about $18 million in 2010). In 1892 he formed a new venture, the

was not a problem, as he then turned his attention to the electric elevator and organized the Sprague Electric Elevator Company in 1892, and with Charles R. Pratt, he developed the Sprague-Pratt Electric Elevator. After building 584 elevators for the tallest buildings and the largest installation—a one-half-million-dollar, forty-nine-car contract with the Central London Tube Railway—Sprague sold his business to the Otis Elevator Company.

The development of the electric railways was momentous. It had a major impact on individual urban transportation by replacing the slow, noisy, smelly, and costly horsepower-based system of streetcars.[77] But it did not reduce the traffic jams in the big cities where both horse-powered transportation and electrically powered transportation often created gridlock.

War of the wires

This took place in capitalist America where private enterprises fought each other for domination of the markets. For example in Cincinnati this competition led to the "war of the wires" concerning the use of single-overhead wire or double-overhead wire systems for supplying electricity to the streetcars. The telephone companies complained that a loud buzzing sound on phone lines was caused by electric interference, and a lawsuit followed. Judge William Howard Taft (later president of the United States) ruled against the single-wire system, but his decision was later overruled. Without going into legal details, the following illustrates the context for the development of applications such as the electric streetcar replacing the horse-powered streetcar.

> *Efforts to consolidate all the Cincinnati area's street railways had been under way since the 1860s. John and Charles Kilgour, brothers who had inherited comfortable fortunes and proposed to increase them through investments in real estate, railroads, banking, and street railways, viewed a monopoly as the natural state of affairs in the street railway business. Charles, who became celebrated as*

Sprague Electric Elevator Company. Sprague and associate Charles Pratt invented the Sprague-Pratt Electric Elevator, which employed Sprague's all-important method of returning power to main supply systems. The elevator had the ability to carry heavier loads and move more quickly than hydraulic or steam elevators. He sold nearly six hundred elevators for buildings around the world before selling the firm to the Otis Elevator Company in 1895. Source: Lemelson-MIT: Inventor Archive. http://web.mit.edu/invent/iow/sprague.html (Accessed December 2014)

[77] In 1880 about 100,000 horses and mules were powering 19,000 streetcars on 3,000 miles of track in three hundred US cities. (Passer, 1972). A horse car required one team in the summer and two in the winter, and the average animal wore itself out in four years. Horses annually ate their value in feed, and companies needed large stables, including a small army of stable hands, several blacksmiths, and a veterinarian (Nye, 1990).

Cincinnati's millionaire bachelor, took over the Franklin Bank, founded by his father, and became a pioneer in the transit business, in which his brother joined him about 1860. John became vice president of the Consolidated Street Railroad Company in 1873. This firm succeeded in taking over many of the area horse car lines.

There were several independents whose stubborn existence frustrated the brothers' master plan, but one by one they picked these off, forming a larger combine, the Cincinnati Street Railway, in 1880…Meanwhile, [his brother] Charles had become intrigued with telegraphy and formed the City and Suburban Telegraph Association in 1873. Its first line connected his office/home with a machine shop he had taken over as an investment. Ever ready to expand his empire, Kilgour added the newly introduced telephone to the telegraph operations in 1877…

By 1890 the Kilgours had acquired control of most transit operations in the metropolitan area. One remaining independent, the Mt. Auburn Cable Railway, was a minor four-mile operation that carried only about five thousand passengers a day. It verged on bankruptcy for much of its history and posed no serious competition to the expanding CSR. Another independent, the Cincinnati Inclined Plane Company (CIP), was more troublesome. In January 1888 its principal stockholder, George A. Smith, had died. The company was already in financial trouble, but the Kilgours waited for the stock to drop even lower. Their hesitation proved fatal. An outside syndicate from Louisville suddenly appeared and bought it out from under their very noses. The new owners were both aggressive and progressive. Within months of their impertinent takeover, they had signed a contract with Frank J. Sprague to electrify the road. The old horsecar line was equipped with electric cars and began running from the city center to the zoo in early June 1889…

This was the first large-scale electric trolley operation in the city. The CIP announced plans to double the size of its system, which used the conventional single trolley wire and a ground return…not long after the new electrics began rolling along Auburn Avenue, the telephone company—Charles Kilgour's City and Suburban Telegraph Association—began to receive complaints about a loud buzzing sound on the phone line. The company assumed that the source of the problem was the single-wire-and-ground-return electric railway, and it asked the courts to intervene…

The Kilgour case ended up in the Cincinnati Superior Court…Taft was overruled, and the CIP was free to operate as it wished…Kilgour, whom a contemporary described as "a ruthless man who overrode all obstacles in his path," seemed unable to admit in public that he was wrong, and he stood by the arguments against the industry-standard single-overhead system. Of course, he had little real need to explain his reasons, because he was not only president and general manager of the CSR; he and his brother Charles, plus a few cronies such as George N. Stone and George Bullock, also owned most of the railway's stock (White, 2005, pp. 377-381).

The work of Sprague and others resulted in the electric tramways, which consumed the electricity generated by the emerging central stations during the daytime. The central stations had originally been formed to supply power for lighting home and offices (mostly when it was dark), and they had no clients for their daytime supply of electricity. The electric streetcars were the ideal clients to use that supply of electrical power.

Stationary applications of DC motors

The Sprague Electric Railway and Motor Company, created by Frank Julian Sprague, was manufacturing the best medium-sized DC motor of the time. Others started to develop DC motors for specific applications such as lathes, ventilation fans, sewing machines, and dental engines. In 1887 the United States had about fifteen manufactures of electric DC motors, and by 1887 more than ten thousand units had been produced.[78] The most important of these companies was the Curtis, Crocker, Wheeler Company that produced six-volt and one-hundred-volt DC motors for sewing machines and other small devices.

Curtis, Crocker, Wheeler Companies

The story of this entrepreneurial business is related to three persons. *Charles Gordon Curtis* (1860–1953) was born in Boston, Massachusetts, and graduated from Columbia University in 1881 as a civil engineer. He also obtained a law degree at New York Law School and then started his own patent law firm. *Schuyler Skaats Wheeler* (1860–1923) was born in New York and also studied at Columbia University. At the age of twenty-one, he left college and became an assistant electrician for the Jablochkoff Light Company. Later he joined Edison's engineering staff (in 1882) where he was one of the builders of the Pearl Street station. *Francis Bacon Crocker* (1861–1921) was born in New York and graduated in 1885 with a PhD degree from Columbia University, where he would later become a professor

---

[78] Data from: Motors and Generators, Industry Report: Motor and Generator Manufacturing. http://business.highbeam.com/industry-reports/equipment/motors-generators

in the newly formed Department of Electrical Engineering. He published papers on electric motors and electric lighting.

Originally Crocker and Curtis partnered in the firm of Curtis & Crocker Electric Motor Co., manufacturing an electric DC motor. The two men took Wheeler into their new partnership in 1886. In 1888 Curtis left the partnership, and Crocker joined forces with his boyhood friend Wheeler to create the Crocker-Wheeler Electric Company in 1889 (dissolving the C & C Electric Motor partnership). Curtis went his own way and formed his own company: Curtis Electric Mfg. Co., a manufacturer of steam turbines. In 1893 the company moved its works from New York City to New Jersey, where it created a new industrial community and called it Ampere (after Andre-Marie Ampère).

Figure 109: 110V DC electric fan by Curtis & Crocker Electric Motor Co. (1887–1890)

The electric lamp acts as the resistance to set the speed of the motor. Changing the lamp changes the speed of the motor.

Source: www.edisontinfoil. com/fans/ccfan.htm

The Crocker-Wheeler Electric Company manufactured small electric DC motors (1/6—1 hp) with additional equipment such as switch boxes that were usable in small electric equipment. They also manufactured an electric fan that ran on 100–110 volts of DC current. Wheeler obtained a patent for a dual-speed switch (US patent №. 460.076, granted on September 22, 1891). Crocker and Wheeler covered the motor design in US patent №. 494.978 (for an electric motor to be used in electric fans, filed on June 20, 1892 and granted on April 4, 1893). They also made dynamos, such as the "ringing generator" (1.000 Hz AC), and battery-charging generators used at telephone companies (Wheeler's US patent №. 503.106, for an armature for dynamo-electric machines and motors, filed on April 24, 1893, and granted on August 8, 1895). These generators were also used in the local and urban DC-distribution systems for electric lighting (Blalock, 2011).

Figure 110: Crocker Wheeler 1/6 H. P. Bipolar Electric DC motor (1891)

Source: www.antiqbuyer.com/All_Archives/OFFICE_ARCHIVE/motorarchive. htm

Diehl Manufacturing Company

Philip H. Diehl (1847–1913) was born in Dalsheim, Germany, and emigrated to New York City in 1868. There he found a job as an apprentice in the Singer Manufacturing Company in New York and the Remington Machine Company in Chicago. He advanced as a machine designer at the

Singer plant in New Jersey (manufacturing sewing machines in large quantities), and he obtained many patents for mechanical improvement in sewing machines.

Around 1886 he developed a variable-speed electric DC motor that could power a sewing machine (US patent №. 356.576, granted January 25, 1887). On April 17, 1888, he was granted US patent №. 381.222 for the adaption of an electric motor for the sewing machine. In addition to the sewing machine motors, he also developed a range of motors for different small applications: the dental motor, for powering dental drilling tools and the ceiling fan motor, for powering ventilation fans (US patent №. 425.995 granted on April 22, 1890). His fan could be powered from a DC system of electricity, nicely fitting in with incandescent lamps. As his ceiling fan was rather a commercial success, he was soon faced with quite a number of competitors.[79] His firm, Diehl & Co., which was founded in 1887, later became Diehl Manufacturing Co. and was purchased by Singer.

Figure 111: Diehl electric motor (1893)
Source: http://www.liveauctioneers.com/item/2252297

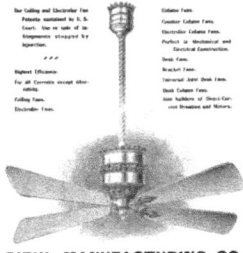

Figure 112: Advertisement for Diehl ceiling fan (1900)
Source: *Electrical World and Engineer*, June 30, 1900. www.afcaforum.com/forum1/21749.html

The breakthrough of the DC motor

Sprague, Curtis & Wheeler, and Diehl—these are just a few of the DC-motor manufacturers that were established after the DC generator was developed. Some of the other manufacturers of DC motors that ran on 110V were: Kendrick & Davis, Ajax Electric Motor Co., Emerson Electric Motor Co., Robbins & Myers. The number of manufacturers illustrates the developing entrepreneurial activities during this burgeoning era of power. After the introduction of the electric-dynamo, removing the barrier of the limited, complicated, and expensive

---

[79] In the warm climate of the southern States, ventilation was the only method of cooling. So mechanically powered systems, using a system of belts, were used—mainly in factories. Also, water-powered ventilation fans where used where running water was available. The applications of another power source (i.e., the electric motor) was a logical one, and its advantages soon made it popular. This created new markets as fans became available in versions that were practical for home use.

use of electrochemical batteries, the *era of light* had started. The application of arc light, followed by the incandescent light, created enormous markets for the supply of electric energy. This resulted in the development of various lighting systems and electric power systems, all dominantly based on the DC current.

When the (often locally based) distribution systems became more and more available, it gave rise to a totally new range of applications in addition to lighting. DC-motor-based applications included powering the washing machine (Figure 113) (early 1900), powering the vapor-compression-based refrigerator, and powering the sewing machine (Figure 114). Applications also included the simple but effective electric fan that helped ventilate houses, offices, shops, restaurants, and factories and the more demanding applications of electric elevators in hotels and offices.

In addition to these applications, a range of industrial applications developed, like the small-motor-based applications of wood- and metalworking and dental tools.

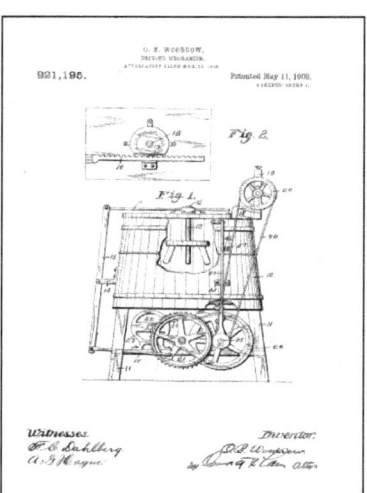

Figure 113: US-patent 921.195 for an electric motor powered washing machine (1909)
Source: USPTO

The electric motor slowly penetrated the printing, textile, and metalworking industries. The originally steam-powered "line and shaft systems" for distribution of rotative power to the individual machines, were becoming powered by electric motors. But their penetration in industrial applications was not impressive.

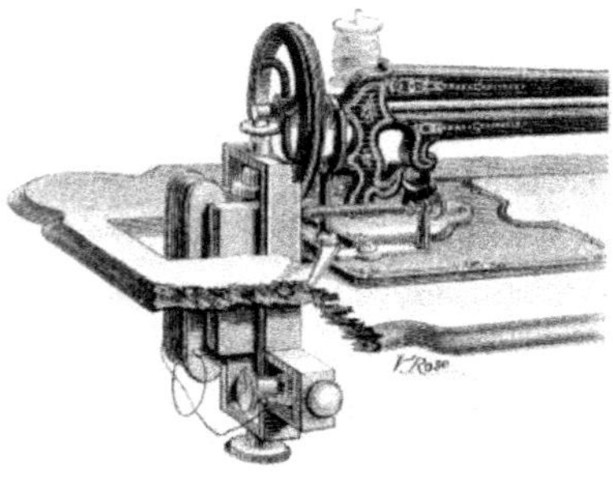

Figure 114: Singer sewing machine with DC motor
Source: http://seaus.free.fr/spip.php?article500

By late 1886, 250 Sprague motors of 0.5 to 15 hp capacity were operating in a number of cities across the United States; in 1889, total electric-motor capacity in manufacturing exceeded 15,000 hp, with over one-quarter of this capacity in printing and publishing establishments. By the early 1890s then, DC motors had become common in manufacturing, but were far from universal (Devine, 1983, p. 355).

The *era of light* became possible after the electric dynamo became available as a source of electric energy; the electricity distribution infrastructure that resulted from it created new opportunities for power applications. Now the *era of power* complemented the earlier era of light. And the electromotive engine—in this case the DC motor—had proved itself useful. It took a while though for the DC motor to emerge as a serious development. Decades had passed since the early days of the development of the DC motor.

Neither the era of power nor the era of light that emerged from all these new applications developed easily. For the entrepreneurial inventors trying to exploit the fruits of their inventive activity, it was a highly competitive world. They had to participate in a battle to defend their work by filing patents—and that could only result in costly patent wars (for example the patent war related to the electric incandescent lamp) (Shaver, 2012).

The invention of the electric induction motor

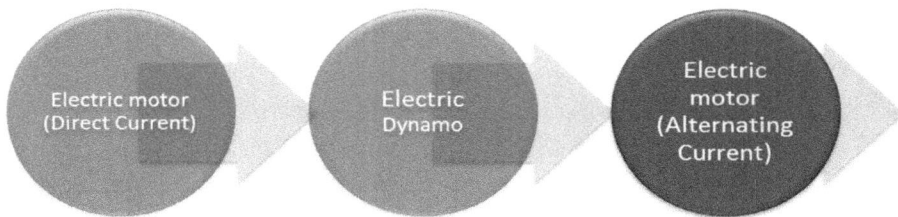

In the preceding part of this case study, the influence of the development of electric light—both the arc lamp and the incandescent lamp—on the development of the magneto-electric dynamo has been explained. It was the symbiosis of the *generation of electricity* by dynamos and the *consumption of that electricity* in the application of electric lighting that gave enormous impetus to the development of electric power systems. These systems consisted of electric dynamos generating electricity, their distribution network, and the "loads" (mostly lamps) consuming the electricity. This electricity was available in a lot of nonstandardized variations (in terms of different voltages, frequencies, lamp- and dynamo-specifications).

To review the origin of electric systems, they started simply and locally with *one-to-one distribution systems* (i.e., arc light in lighthouse with generator, see page 96) and *one-to-few distribution systems* (i.e., block lighting with a generator used in hotels, restaurants, and theaters) were developed. But there were some aspects that hindered further development—for example the inherent problematic character of the arc light (intensity, noisy, smelly, poor reliability, high maintenance) and the use of direct current (DC), which limited the physical expansion of distribution networks. DC distribution networks were limited in their range, due to the fact that the

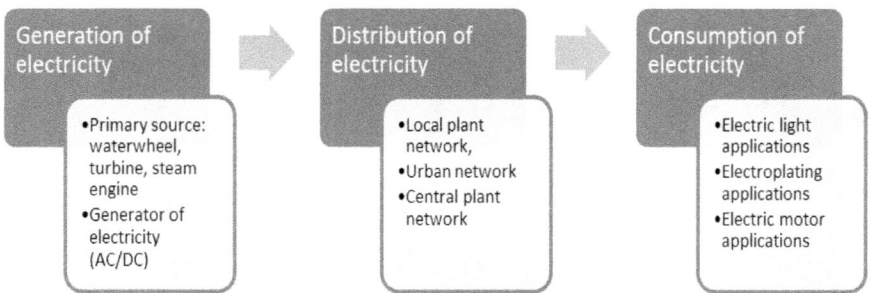

Figure 115: Overview of the early DC-electric systems of generating, distributing, and consumption of electricity

voltage drop in the copper cables became problematic over longer distances. So, a typical *urban distribution system* would use DC to cover an area of about a square mile.

> *A lot of efforts were put in place to create DC networks operating over a larger distance. In 1883 Mr. Charles F. Brush...put upon the market a system in which continuous-current generators of high potential fed storage batteries grouped in series connection. The batteries, in turn, fed, locally, incandescent lamps in houses, hotels, and other buildings. High hopes were held for the success of this plan, but battery difficulties and the dangers incident to the high-potential series system finally caused its downfall. Dr. Weston, Prof. Elihu Thomson, and others attempted, at about this time, to operate a series multiple system in which the potential upon each multiple arrangement of lamps was governed by an automatic device that cut into circuit, compensating resistances whenever the lamps were turned out. This system was introduced in many places, and was especially successful wherever the lighting could be watched and guarded by a local attendant, as in railway stations, department stores, and other semipublic places; but no general solution of the problem was found, and the art hesitated, and its progress was delayed, for the reason that conservative and far-sighted engineers did not recommend its introduction for the general distribution of light and power* (Stanley, 1912, p. 562).

So, DC-based arc-lighting systems had specific characteristics with problems of their own. As mentioned before, the basic problems of the arc light were solved by the new incandescent lamp. This resulted in the development of a range of, manufacturer dependent, incandescent lighting systems. These were systems that had their own limitations, and in their totality they created the "era of light." Then the expansion and further progress more or less halted because the DC motor had only found limited applications—like transport applications, as proved by Sprague's motor for street tramways and elevators and the individual machine tools and appliances that Diehl, Crocker & Wheeler, et al., explored. The further and broader implementation of electricity in society waited for two breakthroughs. One was better distribution, and the other was increased rotative power. The problem was how to transport electricity over distances. And the world was waiting for a self-starting, high-torque motor, available in a range of different versions.

The AC-induction motor, to be described in the following section, is only a part of the developments within the total AC-electric power systems. Firstly there was the development of the generation of electricity by the electric dynamo—the *AC generator* powered by its primary source. Secondly there were developments in the transportation and distribution of AC electricity over electricity networks: here the *AC transformer* was important. And the final development is related to the consumption of electricity by the *AC electric motor*. It would be the AC electric motor, used in "rotative" power applications, that created the breakthrough in the "power applications."

This illustrates the struggling development of the electric systems in the early days of electricity. A development as a whole system from generating electricity, distributing it and finally to consuming electricity. But that all changed when the alternating current, a basic form of electricity that had already been known and used for a long time, again came under the attention of scientists, engineers, and entrepreneurs in the second half of the nineteenth century. Alternating current was to be the breakthrough for electricity to be applied on a large scale.

Alternating current: electromagnetic components

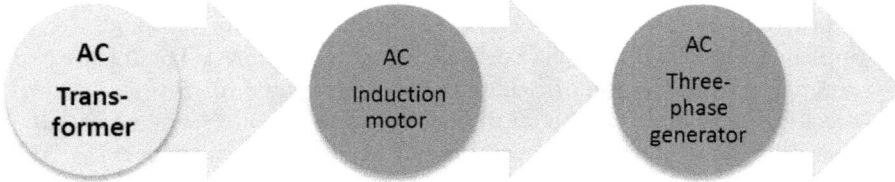

Alternating current (AC) had a big advantage over DC: it could quite easily be used at higher voltages. And, therefore, it could be distributed over large distances (tens of miles). The question was "How could higher voltages than the voltage supplied by the AC generator be created?" Could the voltage of the generator be increased? That was problematic due to technical constraints (e.g. sparking). Could the high voltage be run into offices and residences? That could be dangerous, as Edison would point out so clearly in the "battle of the currents." The answer proved to be the "AC transformer" that could be used to transform AC electricity from a low voltage to a higher voltage (and vice versa).

From induction coil to three-phase transformer

Faraday had already demonstrated in 1831 (with his "ring transformer") that AC current could be transformed using electromagnetism. This was called the "induction coil." But several more inventors who worked on further developing the induction coil were needed before there was a practical version of the "transformer." It was, for example, Pavel Jablochkov who applied a set of coils to his lighting system in Paris in 1876; he connected the primary winding to a source of AC, and the secondary windings were connected to his electric candles. In England Sebastian de Ferranti (together with William Thomson) designed an AC power system in the 1880 to 1882 time period that applied a kind of a transformer. The engineer Frenchman Lucien Gaulard and the English businessman John Dixon Gibbs built a step-down transformer in 1882 that was followed by a step-up transformer. They were granted British patent №. 4.362 in 1882 and №. 1.020 in 1883 for their apparatus. These inventors presented the system first in 1883 at a small electrical exhibition in the Westminster Aquarium in London. Their apparatus had an open iron core, and pushing and pulling the iron core into and out of the coil controlled the voltage.

> *The Gaulard-Gibbs system was subsequently exhibited at the Turin Fair in 1884. It happened that a young Hungarian engineer, Otto T. Bláthy from the Electrical Department of the Ganz Works in Budapest was present at the fair,*

since Ganz was also among the exhibitors…Returning from the fair, Bláthy reported his experiences to the head of Ganz's Electrical Department, K. Zipernowsky, who was permanently experimenting in company of M. Déri. Bláthy's report gave new impetus to the research in progress at Ganz on the subject of the "subdivision of electric light." [It was the work of] three engineers Zipernowsky, Déri, and Bláthy that rapid development of electrical engineering became possible through the realization of the transformer system in 1884–85. The first patent applications were filed on January 2, March 3 (Austria-Hungary), February 18, and March 6, 1885, (Germany) (Asztalos, 1986, pp. 6-9).

George Westinghouse bought the rights to the Gaulard and Gibbs transformer design, and it was Stanley who improved on their design. Since their transformer was not very efficient to produce, it was not very successful. The ZDB transformer exhibited at the Budapest Exhibition in 1885 was a different story (Figure 116).

Figure 116: ZDB transformer (1885)
Source: Wikimedia Commons

In 1885 Ganz & Company exhibited at a local fair in Hungary a system employing alternators wound for a constant and high potential, induction coils connected in parallel arc by their primary circuits and wound with short and, therefore, low potential secondaries, to which lamps were connected. In fact, they disclosed the alternating current system as we now use it. Their transformers were made with closed magnetic circuits, were intelligently designed, and were properly constructed…In October 1885, Zipernowski, Deri & Blathy published a description of their system in the English Electrical Review (Stanley, 1912, p. 564).

Following the above exhibitions, orders poured to the Ganz Works, and the transformer №. 86 was completed in the very year 1885. The hundredth transformer was delivered on March 18, 1886, the thousandth in 1889, and the ten thousandth in 1899 (Asztalos, 1986, p. 11).

These were the early developments that resulted in an electromagnetic device that could transform electricity of a lower voltage to a higher voltage (and vice versa): the "step-up transformer" and the "step-down transformer." It would become an important component in distribution networks.

The idea of using transformers in a distribution network was picked up by William Stanley, who proved the concept in the Great Barrington Experiment in 1886.

At the north end of the village of Great Barrington was an old, deserted rubber mill. This I leased for a trifling sum, and erected in it a 25-horsepower boiler and engine [steam engine] that I purchased for the purpose...We installed in the town plant of Barrington two 50-light and four 25-light transformers...

Figure 117: William Stanley's First Transformer used in the Barrington-experiment (1885)
Source: Wikimedia Commons.

The transformers in the village lit thirteen stores, two hotels, two doctors' offices, one barbershop, and the telephone and post offices. Tile lamps were of 150-, 50-, and 16-candle-power sizes. The length of the line from the laboratory to the centre of the town was about 4,000 feet...

We first devised and tried out at the Great Barrington laboratory the step-up and step-down transformer system now so generally used for power transmission, the generating electromotive force being transformed from 500 to 3,000 volts and from 3,000 back to 500 volts, and then sent over the line downtown (Stanley, 1912, pp. 570-572).

The experiment proved that the transformation from a low-voltage source (the generator) up to a higher voltage for the transmission, followed by transforming it down to a usable voltage at the user side, was feasible. After Stanley's successful experiment, Westinghouse became convinced and adapted the system.

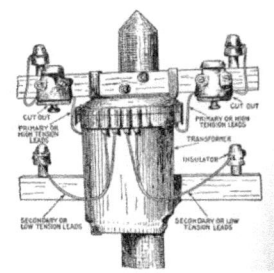

Figure 118: Example of step-down transformer on a pole
Source: Wikimedia Commons.

On April 6, 1886, Messrs. George and H. H. Westinghouse, William Lee Church, Guido Pantalioni, H. H. Jackson, Franklin L. Pope, and Walter C. Kerr came to Great Barringon to visit my laboratory, where they saw the system working for the first time. This visit determined Mr. Westinghouse to actively enter the alternating-current field, as the novelty and scope of the system surprised him greatly (Stanley, 1912, p. 573).

George Westinghouse and William Stanley thus created a transformer that was practical to produce and could be made as a step-up or a step-down transformer. And it could easily be applied in a distribution system, by hanging the transformer on a pole. (See Figure 118: Example of step-down transformer on a pole)

It was the introduction of alternating-current (AC) distribution systems using step-up and step-down transformers that would change the application of electricity on a larger scale. Particularly, it was the ability to convert into high voltage that did the trick.

> *To sum up in a few words: Gaulard and Gibbs considered, developed, and demonstrated crudely the general principle of transformation of electrical energy. Deri, Blathy, and Zipernowski of Budapesth early began research in the same direction. Westinghouse took the crude ideas and, with his engineers, worked out a commercial system and revolutionized the electric art* (Prout, 1921, p. 111).

> *Such, then, is a part of the story of the development and growth of the alternating system in this country. It seems a long step from the limitations of 1884 and 1885, when the maximum area served was approximately sixteen square miles, to the present [1912] service obtainable from a central source that will furnish at equal efficiency an area of 400,000 square miles* (Stanley, 1912, p. 580).

So the distribution problem was, in principle, solved. AC was going to have a bright future as it was going to be used both for lighting and power.

The single-phase AC motor already existed, but it had major drawback. It was not self-starting, as a single-phase motor does not produce any torque at standstill. So, next to the applications of light (both arc lamps and incandescent lamps), the applications of power (the motor-driven applications) were quite limited. But that would change with the development of the *self-starting* AC induction motor.

Alternating current: the induction motor

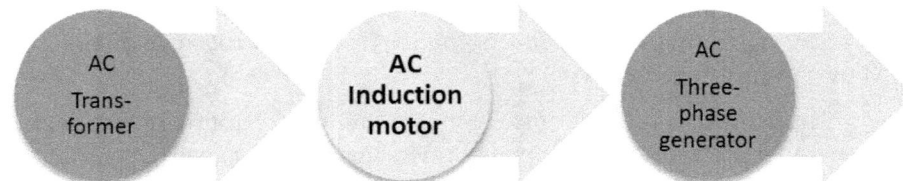

It was many decades after the discoveries around the (single-phase) AC motor and the DC electric motor that the stalled development of the electric motor revived. It was the "induction motor" that overcame a serious deficiency of the alternating-current system—a component that was essential in the so-called "battle of the currents" between systems based on alternating-current (AC) supply and systems based on direct-current (DC) supply.

> *The self-starting characteristic of the induction motor solved the AC motor problem, and it was generally more reliable than DC motors because it had fewer moving parts. Engineers would continue to develop the DC motor into a rugged device for propelling streetcars, elevators, and for other variable-speed applications in industry, where it had an advantage over the constant-speed induction motor. The induction motor became known as the workhorse of industry because it was an integral part of the spreading AC power network that eventually supplanted the DC system* (Ronald Kline, 1987, p. 284).

Science and engineering discover induction

The French physicist François Arago (1786–1853) created the foundation for the induction motor in 1824. He formulated the existence of rotating magnetic fields, termed "Arago's rotations" around 1823–1826 (Figure 119).

> *Francois Arago found that a magnetic needle suspended above a copper disk would rotate as the disk was spun. One year later, Charles Babbage and John Herschel demonstrated the inverse effect: turning a horseshoe magnet beneath a copper disk caused the disk to spin. In either case, no action occurred between the disk and magnet when both were at rest* (Ronald Kline, 1987, p. 285).

This was the basic phenomenon. However, it took a while before this discovery found application in the emerging electric technology. It was just one of the two basic phenomena that were needed for the induction motor.

The other phenomenon was the revolving electrical field.

> *The phenomenon illustrated the conversion of one form of motion (turning a magnet) into another (the rotating disk). And, since electricity, in the form of eddy currents, was only a medium in this conversion, Arago's disk did not demonstrate the action of an electric motor: the conversion of an external source of electricity into motion. The discovery that revealed that these rotations could form the basis of an electric motor was the second phenomenon mentioned above—the production of a revolving magnetic field from an electrical source*

(Ronald Kline, 1987, p. 286).

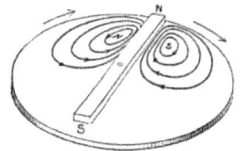

Figure 119: Principle of Arago's rotations between a compass needle and a copper disk

In this schematic explanation of Arago's rotations, turning the disk under the bar magnet induces eddy currents (shown by closed loops) in the disk. The interaction between the magnetic fields of these currents and the magnet causes the magnet to rotate.

Source: R. Mullineux Walmsley, *Modern Practical Electricity* (Chicago 1903), 2:588. (Ronald Kline, 1987)

Other than in a DC motor, which uses a magnetic field fixed in position, the induction motor uses the revolving (or better, rotating) magnetic field. One way to realize this would be to mechanically rotate the permanent magnets. (This is impractical as it begs the question of who or what would move the magnets, but multiphase AC electricity proved to work also.)

It was towards the end of the nineteenth century that several engineers started working on Arago's principle. Among them were the Swede Jonas Wenström (1855–1893), Friedrich Haselwander (1859–1932) of Germany, Charles Bradley (1853–1929) of the United States, the Italian Galileo Ferraris (1847–1897), and the Servian Nikola Tesla (1856–1943). Their work would result in a self-starting AC motor that was developed in a series of steps.

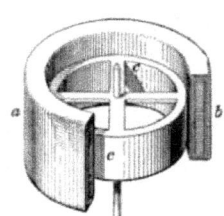

Figure 120: Principle of Arago's rotation and Deprez's rotating filed translated to an induction motor

In this schematic explanation, the out-of-step DC current in the coils (a, b) creates a rotating field that makes the copper disk rotate.

Source: R. Mullineux Walmsley, *Modern Practical Electricity* (Chicago 1903), 2:589.
http://catalog.hathitrust.org/Record/005765090

Step: the principle of the revolving field

The phenomenon of the revolving field was reproduced by Walter Baily in June 1879 in a model exhibited at a meeting at the Royal Society of London (Baily, 1879).

It was the first battery-operated polyphase motor aided by a commutator. This switching device provided the connection between the coils of the electromagnets and the two batteries. Then the Frenchman Marcel Depréz (1843–1918) proved mathematically that a revolving magnetic field could be produced electrically, without the aid of a commutator, with two out-of-step DC currents (actually he used a two-phase DC system that was 180 degrees "out of phase"). He read a paper about his discovery before the Paris Academy of Sciences in 1883 (Ronald Kline, 1987, p. 286). (See Figure 120.)

Step: single/two-phase AC-DC induction motor Next it was Galileo Ferraris who demonstrated a working model of his single-phase DC induction motor in 1885. Nikola Tesla constructed his working two-phase AC induction motor in 1887 and demonstrated it at the American Institute of Electrical Engineers in 1888. He patented his design and was granted US patent №. 382.279, filed November 30, 1887: "A New System of Alternate Current Motors and Transformers." So a two-phase, "out-of-step" DC motor was feasible and did work. It also functioned on alternating current. But the Tesla motor did not live up to its expectations—it was impractical for industrial purposes and needed two-phase AC power. Westinghouse suspended work on it in December 1890 because of financial difficulties within the company (Ronald Kline, 1987, p. 292).

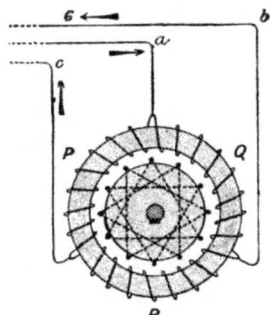

Figure 121: A three-coil armature Q, P, R for a three-phase power supply A, B, C

In this schematic explanation, the three connected coils P, Q, and R—supplied from the generator by a, b, and c—create in the closed coils of the rotor a rotating field that makes the rotor rotate.

Source: R. Mullineux Walmsley, *Modern Practical Electricity* (Chicago 1903), 2:592. http://catalog.hathitrust.org/Record/005765090

Step: the three-phase induction motor So the next question was: why not work with more phases? And why not try an AC power supply that was out of phase? So the idea arose of creating a three-phase AC system with the currents 120 degrees "out of phase" to create a smoother-running induction motor (Figure 121). This idea originated in Europe and became manifest at the International Electrical Exposition at Frankfurt am Main in 1891.[80]

---

[80] One of the objectives of the organizers of the exposition was "to bring forward all decisive materials and arguments concerning "the unpleasant battle of the currents" (Neidhofer, 2007, p. 94).

Step: Closed coils in the armature And last but not least, as the induced currents in the copper armature (the original "Arago disk," later called a copper cylinder, also called "rotor" in these motors) are quite weak, why not apply coils to create electromagnets within the armature? This seemed to have been Tesla's idea.

> *Another, and a very remarkable feature of Mr. Tesla's discovery, was the armature [rotor] with coils closed upon themselves. If we consider for a moment the conditions of the motor described…it will be seen that the armature being merely of soft iron does not exert a mutual attraction upon the field. If the armature be wound with a coil, and a current passed through the latter making it an electromagnet, the power of the motor would be very greatly increased, but this would involve a separate generator. Tesla, however, conceived the plan of winding the armature [rotor] with coils of wire, and closing these coils upon themselves, so that they acted as the secondary of the induction coil for which the two sets of coils became alternately the primary, one set of field coils inducing currents that establish poles to be reacted on by the other*
> (Dood, Leland, & Kline, 1989, p. 1021).

So, in a series of steps, the three-phase AC induction electric motor was developed as part of the total AC electricity distribution system. And in combination with the three-phase AC generator, it became the workhorse of electromotive applications.

Early induction motors

Many people worked on the design of the induction motor. Each of them contributed to the construction of the type of electric motor that would become the workhorse of the years to come: the single/three-phase AC induction motor. In the following paragraphs, their efforts are described in more detail.

Galileo Ferraris induction motor (1885)

The Italian Galileo Ferraris (1847–1897), the son of a pharmacist, went to school in Turin and got his degree in electrical engineering in 1869. He became an assistant at the educational institute *Museo Industriale of Turin* (later the University of Turin), where he taught technical physics from 1877 on. In 1882 he established the School of

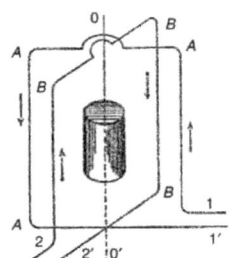

Figure 122: Principle of Ferraris's two-coil AC induction motor (1885, 1888)

In this schematic explanation, the two coils A and B, placed perpendicular to each other, produce a steady revolving field that makes the copper cylinder rotate.

Source: (Neidhofer, 2007)

Electrotechnology with Laboratory at the *Museo Industriale*. By 1884, Italy—still a young country having only been united since 1861—wanted its own international exhibition. This was held at Turin, which was the first capital of Italy. Ferraris was made president of the electrical department of the exhibition.

In 1885 Ferraris, then professor of technical physics at the University of Turin, devised an apparatus intended for classroom demonstrations of Arago's disk. The result was an early induction motor with two coils (Figure 122, Figure 123), arranged perpendicular to each other and fed by two alternating currents of the same amplitude and frequency but with a phase displacement of 1/4 period (also indicated as "0° out of phase").

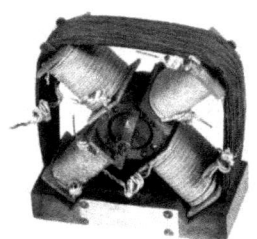

Figure 123: Ferraris's induction motor: fourth prototype (1886)
Source: Physics Museum of Sardina, http://www.webalice.it/sergio.arienti/immvite4/motoreeferRARIS_cmr_4.gif

At the time of the Turin exhibition, electricity was used almost exclusively for lighting, but people were beginning to think about electric motors. The idea that a rotating magnetic field might cause a suitable "rotor" to revolve was not new. Walter Baily, for example, had exhibited in London in 1879 a device in which two sets of electromagnets were switched alternatively causing a copper disc to rotate. Ferraris's transformer studies led him to consider the fact that the primary and secondary currents were out of phase. In the summer of 1885 he conceived the idea that two out-of-phase, but synchronized, currents might be used to produce two magnetic fields that could be combined to produce a rotating field without any need for switching or for moving parts. This idea, which is commonplace to electrical engineers now, was a complete novelty in the 1880s. Ferraris published it in a paper to the Royal Academy of Sciences in Turin in 1888. This was quickly translated into English, and published in the journal Industries, later the same year. In "Il Nuovo Cimento," Ferraris published eleven papers: ten on electromagnetism and one on electricity.[81]

The message spread like wildfire but, at the 1889 World Exposition in Paris, France, Ferraris had to recognize that other researchers, Nikola Tesla in particular, had similar ideas (Neidhofer, 2007, p. 89).

---

[81] Source of text: http://incredible-people.com/biographies/galileo-ferraris/. No author.

Ferraris, being a scientist and not an entrepreneur, did not apply for patents. It was the Westinghouse Company who persuaded him to consider applying for an American patent. He led many organizations in Italy and was regarded as the foremost authority in electricity in Italy. One of his students, Guido Pantaleoni, went on to work for Westinghouse and to be an instrumental link for licensing patents from Europe. His work later became the subject of a patent claims dispute by Nikola Tesla, working for Westinghouse, that Tesla's work was prior to Ferraris's work.

Charles Schenk Bradley (1888)

In the United States, Charles Schenk Bradley (1853–1929) worked for Edison around 1880, before he created his own laboratory in 1883. There he developed the idea for an electric heating stove (US design patent №. D16519, February 6, 1886) with polyphase AC systems. In 1887 he built his three-phase generator and was granted US patent №. 390.439

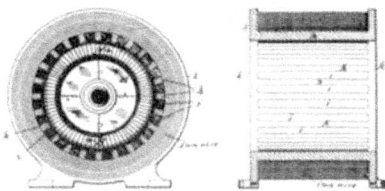

Figure 124: Bradley's induction motor
Source: USPTO

Table 14: Patents by Charles Bradley relating to the AC induction motor, generator, and distribution system

Patent №	Granted	Description
US 390.439	October 2, 1888	Dynamo-electric machine: Two-phase AC motor with 90^0 phase shift (filed May 9, 1887)
US 394.818	December 18, 1888	Dynamo-electric machinery (filed July 25, 1887)
US 404.465	June 4, 1889	Electric motor: Two-phase AC motor (filed October 5, 1888)
US 404.466	June 4, 1889	Electric motor: automatically switched field circuit (filed February 18, 1889)
US 409.450	August 20, 1889	System of electrical distribution: Three-phase system with 120^0 phase shift (filed October 20, 1888)
US 463.852	November 24, 1891	Synchronous telegraph (filed January 28, 1887)
US 438.602	October 21, 1890	Alternating-current generator and motor: construction of self-exiting generators with two magnetic systems (filed June 23, 1890)
US 492.480	February 28, 1893	Transformer and means for developing rotary magnetic fields: converting single-phase current into polyphase current (filed March 17, 1892)
US 508.807	November 14, 1893	Converter system for electric railways: (filed May 31, 1887)
US 514.586	February 13, 1894	Electrical transmission of power: usable for electrically propelled cars or other vehicles

Source: USPTO

on October 2, 1888, for a dynamo-electric machine using a two-phase system ninety degrees out of phase (Figure 124). This was followed by US patent №. 404.465 for a two-phase electric motor using four wires, granted on June 4, 1889. On August 20, 1889, he was granted US patent №. 409.450 for a "System of Electric Distribution." In this patent he described the three-phase system 120 degrees out of phase.

> *In 1887 Bradley was the first to apply for a patent concerning two-phase AC power transmission with four wires. In 1888 he was close to discovering the three-phase AC principle when specifying terminal connections on a ring winding at three symmetrical outer points. Another patent issued in 1888 referred to a two-phase induction motor, actually the first one with a secondary armature completely short-circuited (cage stator)* (Neidhofer, 2007, p. 89).

He did not use his patents to engage in entrepreneurial activities and never put his inventions into practice. Later Bradley also obtained patents for a synchronous telegraph (US patent №. 463.852 of November 24, 1891) and a converter system for an electrical railway (US patent №. 508.807 of November 14, 1893).

The polyphase induction motor

So there it was: the induction motor that would run off AC electricity that was "out of step." As it—generally speaking—could not run on a single AC phase, a polyphase electricity supply was needed. The question now became "how many phases could do the job?" Was it to be a two-phase system, or was it to be a three-phase motor? That is where the champion of the induction motor, Nikola Tesla, came on stage. Working for Westinghouse, he developed a polyphase power system of a two-phase generator and a two-phase induction motor. Meanwhile, in Europe, it was Mikhail Osipovich Dolivo-Dobrowolsky who, with others, developed a three-phase AC power system with a three-phase AC generator and a three-phase AC motor.

Nikola Tesla two-phase induction motor (1887)

The Serbian Nikola Tesla (1856–1943) was a bright young man who passed his tumultuous youth and adolescence in Serbia and Austria. His parents wanted him to become a priest, but that was not to be. He became fascinated by electricity.

> *During all those years my parents never wavered in their resolve to make me embrace the clergy, the mere thought of which filled me with dread. I had become intensely interested in electricity under the stimulating influence of my professor of*

> *physics, who was an ingenious man and often demonstrated the principles by apparatus of his own invention. Among these I recall a device in the shape of a freely rotatable bulb, with tinfoil coatings, which was made to spin rapidly when connected to a static machine. It is impossible for me to convey an adequate idea of the intensity of feeling I experienced in witnessing his exhibitions of these mysterious phenomena. Every impression produced a thousand echoes in my mind. I wanted to know more of this wonderful force; I longed for experiment and investigation and resigned myself to the inevitable with aching heart* (Tesla, 2007).

His father changed his mind after Tesla recovered from a nine-month sickness from cholera. Tesla was allowed to go to the Polytechnic school in Gratz, followed later by the University in Praque. After working in Budapest, where he worked for the Austro-Hungarian state telephone system, he was offered a position in Paris in 1882 at the French Edison Electric Lift Company. From there he went to the United States in 1884 and worked for Edison's Machine Works. It was there that he met Thomas Edison in person:

> *The meeting with Edison was a memorable event in my life. I was amazed at this wonderful man who, without early advantages and scientific training, had accomplished so much. I had studied a dozen languages, delved in literature and art, and had spent my best years in libraries reading all sorts of stuff that fell into my hands, from Newton's Principia to the novels of Paul de Kock, and felt that most of my life had been squandered. But it did not take long before I recognized that it was the best thing I could have done. Within a few weeks, I had won Edison's confidence, and it came about in this way. The S.S. Oregon, the fastest passenger steamer at that time, had both of its lighting machines disabled, and its sailing was delayed. As the superstructure had been built after their installation, it was impossible to remove them from the hold. The predicament was a serious one, and Edison was much annoyed. In the evening I took the necessary instruments with me and went aboard the vessel where I stayed for the night. The dynamos were in bad condition, having several short-circuits and breaks, but with the assistance of the crew, I succeeded in putting them in good shape. At five o'clock in the morning, when passing along Fifth Avenue on my way to the shop, I met Edison with Batchellor and a few others as they were returning home to retire. "Here is our Parisian running around at night," he said. When I told him that I was coming from the Oregon and had repaired both machines, he looked at me in silence and walked away without another word. But when he had gone*

some distance I heard him remark: "Batchellor, this is a d-n good man," and from that time on, I had full freedom in directing the work (Tesla, 2007).

After proving that he could handle quite difficult problems, Tesla was even offered the task of completely redesigning the Edison Company's direct-current generators. But he did not stay long:

> *For nearly a year, my regular hours were from 10:30 a.m. until five o'clock the next morning without a day's exception. Edison said to me: "I have had many hardworking assistants, but you take the cake." During this period I designed twenty-four different types of standard machines with short cores and of uniform pattern, which replaced the old ones. The manager had promised me fifty-thousand dollars on the completion of this task, but it turned out to be a practical joke. This gave me a painful shock, and I resigned my position* (Tesla, 2007).

In 1884 he created his own company, Tesla Electrical Light & Manufacturing Company, backed by Benjamin A. Vail (lawyer and politician) and Robert Lane (businessman), both keen to enter the promising field of electric lighting. Across the country dozens of businessmen such as Vail and Lane were intrigued by the new electrical industry, and they established new companies to manufacture arc-lighting equipment (Carlson, 2013, p. 74). The company installed electrical arc-light-based illumination systems designed by Tesla. It also had designs for dynamo-electric machine commutators, based on the first patents issued to Tesla in the United States (US Patent №. 334.823 granted on January 26, 1886).

Table 15: Some Tesla patents owned by the Tesla Electric Light and Manufacturing Company

Patent №	Date	Description
US 334.823	January 26, 1886	Commutator for dynamo-electric machines (filed May 6, 1885)
US 335.786	February 9, 1886	Electric arc (filed March 30, 1885)
US 335.787	February 9, 1886	Electric arc (filed July 13, 1885)
US 336.961	March 2, 1886	Regulator for dynamo-electric machines (filed May 18, 1885)
US 336.962	March 2, 1886	Regulator for dynamo-electric machines (filed June 1, 1885)
US 350.954	October 19, 1886	Regulator for dynamo-electric machines (filed January 18, 1886)
US 359.748	March 22, 1887	Dynamo-electric machine (filed January 14, 1886)

Source: USPTO

Tesla proposed that the company should go on to develop his ideas for alternating-current transmission systems and motors. The investors disagreed and eventually fired him, leaving him penniless. Tesla was forced to work as a ditch digger for two dollars per day. Tesla considered the winter of 1886–1887 as a time of "terrible headaches and bitter tears." In April 1887 Tesla started a new company, the Tesla Electric Company, which he remembered as following:

> *Immediately thereafter some people approached me with the proposal of forming an arc light company under my name, to which I agreed. Here finally was an opportunity to develop the motor, but when I broached the subject to my new associates, they said: "No, we want the arc lamp. We don't care for this alternating current of yours." In 1886 my system of arc lighting was perfected and adopted for factory and municipal lighting, and I was free, but with no other possession than a beautifully engraved certificate of stock of hypothetical value. Then followed a period of struggle in the new medium for which I was not fitted, but the reward came in the end, and in April, 1887, the Tesla Electric Company was organized, providing a laboratory and facilities. The motors I built there were exactly as I had imagined them. I made no attempt to improve the design, but merely reproduced the pictures as they appeared to my vision, and the operation was always as I expected* (Tesla, 2007).

Charles F. Peck and Alfred S. Brown financially backed Tesla. Peck was a lawyer already active in the telegraph business with his companions Alfred Brown and John Evans in their company Mutual Union. Brown was a director of the telegraph company Western Union. They were mainly interested in the more promising applications of DC motors, a rapidly developing market at that time where DC was the dominant form of electricity. But Tesla was more interested in AC systems, and he had to convince his investors of AC's potential. He did not want these investors to abandon him as his earlier investors had. So he devised his "Egg of Columbus" demonstration (Figure 125) (Carlson, 2013, pp. 77, 90-92).

Figure 125: Tesla's Egg of Columbus (1893)
Source: Nikola Tesla Museum, Belgrado

He had approached a Wall Street capitalist—a prominent lawyer—with a view of getting financial support, and this gentleman called in a friend of his, a well-known engineer at the head of one of the big corporations in New York, to pass upon the merits of the scheme. This man was a practical expert who knew of the failures in the industrial exploitation of alternating currents and was distinctly prejudiced to a point of not caring even to witness some tests. After several discouraging conferences, Mr. Tesla had an inspiration. Everybody has heard of the "Egg of Columbus." The saying goes that at a certain dinner the great explorer asked some scoffers of his project to balance an egg on its end. They tried it in vain. He then took it and, cracking the shell slightly by a gentle blow, made it stand upright. This may be a myth, but the fact is that he was granted an audience by Isabella, the Queen of Spain, and won her support. There is a suspicion that she was more impressed by his portly bearing than the prospect of his discovery. Whatever it might have been, the queen pawned her jewels and three ships were equipt [sic] for him, and so it happened that the Germans got all that was coming to them in this war. But to return to Tesla's reminiscence. He said to these men, "Do you know the story of the Egg of Columbus?" Of course they did. "Well," he continued, "what if I could make an egg stand on the pointed end without cracking the shell?" "If you could do this, we would admit that you had gone Columbus one better." "And would you be willing to go out of your way as much as Isabella?" "We have no crown jewels to pawn," said the lawyer, who was a wit, "but there are a few ducats in our buckskins, and we might help you to an extent." (Carlson, 2013).

Highly impressed by the demonstration, Peck and Brown became ardent supporters of Tesla's work on AC motors.

It was in late 1887 when Tesla, with the help of patent attorney Parker W. Page—son of Charles Grafton Page—filed for several US patents in the field of polyphase AC generators, power transmission, transformers, motors, and lighting. These patent applications were divided into other patent applications and resulted in a number of patents granted on May 18, 1888. In totality they describe a system for the conversion, transmission, and utilization of power created by means of electrical energy: the generator to convert mechanical power into electrical energy and the motor to convert the electric energy into mechanical power. This process had already been done before in earlier systems, but in this case Tesla applied high voltages, got rid of the commutator by using AC, and applied the newly discovered rotating magnetic field of the induction motor. It was the patents filed in October and November that would be the essential ones.

For Tesla, alternating current was a logical approach to the electric systems. Remember, the commutator was a troublesome device in generator and motor design: it sparked, burned, and failed the system. The polyphase AC motors did not have any of those drawbacks as they used no commutator but instead used slip rings (Hughes, 1993, pp. 115-117).

His work covered the two-phase induction motor that used two-phase alternating currents that were ninety degrees "out of step."

> *[Tesla] constructed in 1888 [a motor that] had primary coils wound on inwardly projecting pole pieces, inside of which rotated an iron-drum secondary covered by closed-circuited copper windings...The primary was fed by two AC currents 90 degrees out of phase (i.e., the second current started when the first reached its maximum value). This two-phase motor worked on the same principles as the one Ferraris built. But Tesla's motor was more powerful and efficient because its secondary windings channeled the eddy currents into paths where they interacted more strongly with the revolving magnetic field...The induction motor ran most efficiently at constant speed and had low starting torque—qualities obviously unsuitable for streetcars. The motor was also impractical for industrial purposes because it operated poorly at the relatively high frequency currents then in use (133 hertz) and required two-phase power (AC generators were then all single phase)* (Ronald Kline, 1987, pp. 290, 291).

Table 16: GB patents by Nikola Tesla concerning the AC induction motor, generator, and distribution system (1888–1891)

Patent №	Date	Description
GB 6.481	June 1, 1888	Improvements relating to the electrical transmission of power and to apparatus therefore (filed May 1, 1888)
GB 6.502	June 1, 1888	Improvements relating to the generation and distribution of electric currents and to apparatus therefore (filed May 1, 1888)
GB 6.527	May 18, 1889	Improvements relating to electric motors (filed April 16, 1889)
GB 16.709	December 7, 1889	Improvements relating to the conversion of alternating into direct electric currents (filed October 22, 1889)
GB 19.420	January 11, 1890	Improvements in alternating current electromagnetic motors (filed December 3, 1889)
GB 19.426	December 3, 1889	Improvements in the construction and mode of operating alternating current motors (filed December 3, 1889)
GB 11.473	August 22, 1891	Improvements in alternating current electromagnetic motors (filed July 6, 1891)

Source: USPTO

Tesla's following work was also patented (Table 16). (These were just a few of the three hundred patents he obtained in his life.) This shows his dedication to the developing systems for the generation, transmission, distribution, and use of AC-electric power—the polyphase systems as they were called then. The polyphase system had a major problem though: it used a lot of copper wires (four or six wires) to distribute the electric energy between generator and motor—and copper was expensive. So Tesla developed an induction motor for a single-phase system—the "split-phase motors"—which he did not intend to patent and did not even mention to his investors (Figure 126). This surprised attorney Parker Page and Brown who recognized the market demand for such motors.

> *The result of this disagreement and confusion with Brown and Page was that Tesla wound up securing two groups of patents: one set covering his ideas for the polyphase, multiwire motors and systems, while the second set covered the more practical split-phase or two-wire motors. It was unfortunate that Tesla delayed filing his split-phase applications because the delay weakened his priority claims and led to patent litigation that lasted for the next fifteen years*
> (Carlson, 2013).

The investors, Peck and Brown, decided to commercialize Tesla's patents instead of starting their own manufacturing companies. So Tesla read a paper at the American Institute for Electrical Engineers (AIEE) in May 1888. That presentation had the result they were looking for: George Westinghouse became interested. It was George Westinghouse who licensed Tesla's patents and employed Tesla for one year as a consultant—for a generous $2,000 per month salary[82]—to develop them. Westinghouse paid Tesla $20,000[83] in cash and $50,000 in Westinghouse stock and also agreed to allow Tesla a royalty of $2.50 per horsepower on all alternating-current capacity that the

Figure 126: Tesla's induction motor (1887)
Source: British Science Museum

---

[82] This income would be equivalent to more than $ 45,000 in 2010, calculated on the basis historic standard of living. Source: Measuring Worth at www.measuringworth.com/uscompare/relativevalue.php.
[83] This amount would be equivalent to more than $ 480,000 in 2010, calculated on the basis historic standard of living. Source: Measuring Worth at www.measuringworth.com/uscompare/relativevalue.php.

company sold. Minimum royalties were $5,000, $10,000, and $15,000 for the first, second, and later years. This was quite a generous deal, maybe too generous. When Westinghouse later ran into financial problems, Tesla agreed to accept a cash settlement of $216,000 in lieu of royalties (Carlson, 2013, p. 119).

The motor development failed, as the company did not succeed in creating a practical induction motor, and Westinghouse abandoned the project.

> *In the early part of 1888, an arrangement was made with the Westinghouse Company for the manufacture of the motors on a large scale. But great difficulties had still to be overcome. My system was based on the use of low frequency currents, and the Westinghouse experts had adopted 133 cycles with the object of securing advantages in the transformation. They did not want to depart from their standard forms of apparatus, and my efforts had to be concentrated upon adapting the motor to these conditions. Another necessity was to produce a motor capable of running efficiently at this frequency on two wires, which was not easy of accomplishment. At the close of 1889, however, my services in Pittsburg being no longer essential, I returned to New York and resumed experimental work in a laboratory on Grand Street, where I began immediately the design of high frequency machines* (Tesla, 2007).

Tesla then started other projects, and the concept of the induction motor was picked up by other companies like the *Maschinefabrik Oerlikon* of Zurich and *Allgemeine Elektricttats Gesellschaft* (AEG) of Berlin. It was there that Michael von Dolivo-Dobrowsky and Charles Brown succeeded in building successful three-phase electric induction motors (Ronald Kline, 1987, p. 293).

Dobrowolsky's three-phase induction motor (1888)

The Russian Mikhail Osipovich Dolivo-Dobrowolsky (1862–1919) came from a mixed heritage: a Polish noble family originating from the polish region of Mazowsze and a Russian noble family. He studied at the Faculty of Chemistry of the University of Riga. German was the language of instruction in this private university, and one of the largest groups of ethnic Poles were from the lands of present Latvia, Estonia, and Lithuania. On March 13, 1881, Tsar Alexander II was the victim of an assassination plot in St. Petersburg. After that event all suspected students of Polish origin were relegated from the Russian higher education institutions. One of the students expelled was Mikhail Dolivo-Dobrowolsky (Ciok, 2009). [84]

So Dobrowolsky emigrated to Germany and studied electrical engineering at the Darmstadt University of Technology. He finished his studies in 1884. Next he worked at the Deutsche *Edison-Gesellschaft für angewandte Elektricität-DEG* (later in 1887 for AEG, *Allgemeine Elektricttats Gesellschaft*, originally the German Edison Company) where he continued his work based on the Ferraris concept and the experiences of the Tesla induction motor. In 1888 he constructed a three-phase alternating current motor with a rotating magnet.

> *Dobrowolsky and Brown succeeded where Tesla and Scott had failed mainly on the basis of more thorough engineering research and development. Dobrowolsky began his work in mid-1888 on learning of the research of Tesla and Ferraris from the technical journals. His first step was to question Ferraris's theory that an induction motor's maximum efficiency was 50 percent. Relying on his*

Table 17: Early patents by Dolivo-Dobrowolsky for the AC induction motor, generator, and distribution system

Patent №	Granted	Description
US 422.746	March 4, 1890	Electrical induction apparatus or transformer (filed January 8, 1890).
US 427.978	May 13, 1890	Alternating current motor (filed November 13, 1889)—Equivalent to other European patents like German patent 51.083 dd. March 8, 1889
US 455.683	July 7, 1891	Transmission of alternating currents of different phase (filed March 28, 1891)
US 456.804	July 28, 1891	Alternating current motor (filed on December 23, 1890)
US 469.515	February 23, 1892	Electric machine: AC motor (filed August 3, 1891)
US 503.038	August 8, 1893	Regulation of alternating current motors (filed December 23, 1890)—Equivalent to other European patents like Great Britain Patent 20.425 dd April 14, 1891
US 540.153	May 28, 1895	Apparatus for determining differences between phases of two electrical currents (filed October 27, 1893)—Equivalent to other European patents like German patent 68.215 dd. April 14, 1892
US 549.449	November 5, 1895	Apparatus for indicating difference of phase (filed April 20, 1890)— Equivalent to other European patents like Great Britain patent 23.113 of December 16, 1892, German patent 69.159 of April 14, 1892

Source: USPTO

---

[84] Text translated from Polish source: 'Michał Doliwo-Dobrowolski - współtwórca cywilizacji technicznej XX wieku.' (Michał Doliwo-Dobrowolski—cocreator of XX century technical civilization).

considerable knowledge of DC machines, Dobrowolsky built an experimental induction motor in the fall of 1888 with a "squirrel-cage" secondary. This secondary, which he patented in 1889, consisted of copper bars laid in slots near its outer periphery...In 1889, he designed a 1/10-HP motor along these lines that had an efficiency of 80 percent and a relatively high starting torque...He then built a motor with an improved primary winding and a rotating secondary connected by slip rings on its shaft to external, lever-operated rheostats. The new primary winding decreased magnetic leakage, and the rheostats, which he patented in 1891, enabled him to increase secondary resistance at starting (Ronald Kline, 1987, p. 293).

Dobrowolsky obtained several patents related to components of an AC system—for example a transformer for which he was granted German patent №. 56.359 on August 29, 1889. He was later granted an US patent for an "electrical induction apparatus or transformer," an AC transformer (US patent №. 422.746 granted on March 4, 1890). For a design of an AC motor, he was granted German patent №. 51.083 on March 8, 1889. The same design was granted US patent №. 427.978 on May 13, 1890, for an "alternating current motor." This design was already patented in 1889 in other European countries such as France, Belgium, Luxemburg, Italy, and Switzerland. Dobrowolsky was also granted a US patent №. 455.683 for the "transmission of alternating currents of different phase" on July 7,

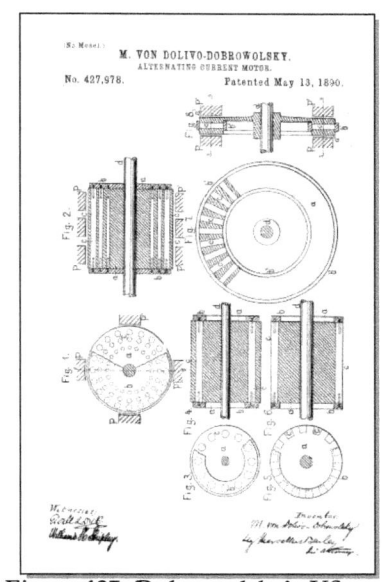

Figure 127: Dobrowolsky's US patent 427.978 for the alternating current motor (1890)
Source: USPTO

1891. Later he added other devices to the systems when he was granted US patent №. 503.038 on August 8, 1893, for the "regulation of alternating electric current motors." This subject was also already patented earlier in numerous other countries such as Belgium (1890), Luxemburg (1890), Italy (1891), France (1891), England (1891), Spain (1891), Austria-Hungary (1891), and Switzerland (1892). So Dobrowolsky's invention was internationally and widely covered by patents (Table 17).

For the International Electro-Technical Exposition at Frankfurt in 1891, Dobrowolsky designed the three-phase generator and transmission system. He additionally developed, in cooperation with Charles Brown, the spectacular induction motors. So it was not just that Dobrowolsky developed the components of the electrical system, he developed the whole three-phase AC system that was exhibited at the Exhibition in 1891. It was a system for the generation, distribution, and use of electricity by induction "engines."

> *In the summer of 1890, AEG and Oerlikon signed a licensing contract on patents and agreed to exchange the results of polyphase research. After viewing Dobrowolsky's motors that summer, Brown returned to Oerlikon and experimented on the best types of primary and secondary windings. Within a few weeks, he built a 1- to 2-HP motor with two important improvements: a primary winding of copper wires laid in many slots around the inner periphery of its core, and a squirrel-cage secondary with bars insulated from its iron core...This primary winding further reduced magnetic leakage, and insulating the secondary bars provided a more conductive path for eddy currents. Later in 1890, Dobrowolsky reversed this design and built a 2- to 3-HP motor with a revolving primary. Its stationary secondary consisted of copper bars laid in slots around the inner periphery of its core and was connected to three starting rheostats. Dobrowolsky exhibited this motor, which had an efficiency of 80 percent, to the Berlin Elektrotechnischer Verein in March 1891. He designed the 100-HP motor shown at the Frankfurt Exhibition later that year by "scaling up" this model...Near the end of the exhibition, Brown displayed the 20-HP motor, which he had scaled up from his 1- to 2-HP model* (Ronald Kline, 1987, p. 294).

The International Electrical Exhibition at Frankfurt (1891)

As described, the early development of the polyphase induction motor was taking place in a lot of countries. In Europe it culminated at the International Electrical Exhibition that was held in 1891 at Frankfurt am Main.

> *In 1891 European firms displayed polyphase induction motors of 20 and 100 HP at the International Electrical Exhibition in Frankfurt, Germany. Engineering attention was drawn to these motors because they operated at the terminus of the most impressive electrical transmission line to date. Three-phase current at 15,000 volts (later 25,000) flowed 175 kilometers from hydro-electric generators at Lauffen am Neckar to the exhibition building at Frankfurt am*

Main, much further than any previous transmission...The engineers responsible for the motors were the leading designers of the two firms that supplied the equipment for the Lauffen-Frankfurt system: the Maschinenfabrik Oerlikon of Zurich and the Allgemeine Elektrizitits-Gesellschaft (AEG) of Berlin...Brown, the designer of the 20-HP motor, was technical director of the electrical section at Oerlikon...Brown designed the polyphase generators for the Frankfurt system, whose revolving-field construction became standard throughout the world. Michael von Dolivo-Dobrowolsky, [was] the designer of the 100-HP motor...The motors of Dobrowolsky and Brown, rather than those of Tesla and Scott, became the prototypes of subsequent induction motors in Europe and the United States (Ronald Kline, 1987, pp. 292-294).

Figure 128: Poster for the International Electrical Exhibition in Frankfurt (1891)

Source: http://www.vde.com/wiki/chronik_neu/Wiki-eiten/Ausstellungen_und_Museen.aspx

The international exposition ended with a one-week Electrical Congress, where Dolivo-Dobrowolsky read the paper *Electrical Transmission of Power by Alternating Currents*. He concluded: "With reference to the unsynchronous, especially the so-called rotary current (drehstrom) motors, which were first suggested and used by Tesla and Ferraris, I believe, I am able to state that I have succeeded in making them very economical and practical." An American speaker, Ludwig Guttman, was skeptical about the benefit of three-phase AC and stated "New types of motors, unknown up to now, are progressing in development, requiring not more than the common two-line system."

Figure 129: Drobrowolsky's 100 HP motor displayed at the Frankfurt Electrical Exhibition (1891)

Source: *Offizieller Bericht über die Internationale Elektrotechnische Ausstellung in Frankfurt am Main*, 1891, 2 vols. [Frankfurt 1893–94], 1:387) (Ronald Kline, 1987)

Whatever the different view on the future, the great event in 1891 went down in history as the birth of long-distance, three-phase power transmission. In 1893 the *London Electrical Review* remarked that:

> *It is astonishing how little has been done commercially with the two-phase system, which was so much written up and introduced about five years ago. The Tesla system of motors seems to be entirely eclipsed by the later three-phase motors [of Dobrowolsky and Brown]* (Ronald Kline, 1987, p. 296).

Parallel development of the AC generator

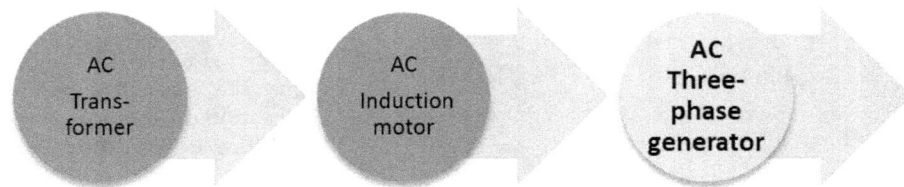

In the preceding pages, the development of the (polyphase) alternating current induction motor is described. These motors could only function when supplied with (polyphase) alternating-current electricity. The survivor of the different versions of the polyphase induction motors proved to be the three-phase AC induction motor. That type of electricity had to be generated by special dynamos: the three-phase AC generators. So the development of three-phase AC generator runs parallel to that of the three-phase AC induction motor. This is illustrated by the following events:

Charles Bradley built his three-phase generator in 1887 and was granted US patent №. 390.439 on October 2, 1888, for a dynamo electric machine using a two-phase system ninety degrees out of phase.

Nikola Tesla designed his two-phase system and designed, in addition to the two-phase induction motor, a two-phase generator. On May 1, 1888, he was granted US patent № 382.280: Electrical transmission of power: two-phase AC induction motor in combination with a two-phase AC dynamo (filed October 12, 1887).

Michael Dolivo-Dobrowolsky designed the revolutionary three-phase generator that was used at the Lauffen project to feed electricity for the Electro-Technical Exposition at Frankfurt in 1891.

All these developments took place around the same time frame (1887–1890) and became well known. However, some others participants in the birth of the three-phase AC system are not so well known, due to a range of circumstances. They disappeared from history over time. Some of their (technical) contributions did not result in an (economic) impact. Others did not survive when they had to fight the already dominant players in the market, or other (often large) companies absorbed their activities. In other words, for a multitude of reasons, one can be sure that behind any well-known contributors, there were numerous unknown contributors.

Friedrich August Haselwander (1887)[85]

The German Friedrich August Haselwander (1859–1932) studied at the *Technische Hochschule Karlsruhe* and the universities in Strasburg and Munich. After his military service, he returned to Offenburg in 1884 and became an independent electrician. His first patent in 1880 was for an arc light. In 1887 he installed his first generator at the works of the Adrion Company. Thus, Haselwander is credited with having built the first three-phase synchronous generator with salient poles in 1887. Furthermore, he recognized the principle of three-phase power transmission with two synchronous machines and three transmission wires, a system he presented in 1888 at a world premiere.

Figure 130: Haselwander's three-phase AC generator (1887) (*Drehstrom maschine*)
Source: Wikimedia Commons

Haselwander was one of those engineers who did not make it. His application for a patent was disputed, and he could not finance the following patent dispute (then expected to be 30 million Deutch Marks).

> *Haselwander meldete seine Erfindung zum Patent an, hatte dabei aber wenig Glück: zuerst verschlampte sein Anwalt die Anmeldung, dann wurde sie wegen angeblicher Unklarheiten in der Beschreibung zurückgewiesen. Wahrscheinlich hatte der Prüfer beim Patentamt in Berlin Haselwanders Ideen einfach nicht richtig verstanden oder er hatte die Tragweite der Erfindung verkannt. Erst 1889—zwei Jahre später—bekam Haselwander das Patent zuerkannt. Es hagelte aber Einsprüche und Haselwander fehlte das "Kleingeld" um sich auf einen langen Patentstreit (der Streitwert lag bei 30 Millionen Mark) gegen die vermögende Großindustrie einzulassen…Die Großindustrie wie*

---

[85] Source: information found at: Landesarchiv Baden-Wuertenberg. http://www2.landesarchiv-bw.de/ofs21/olf/ einfueh.php?bestand=14952, http://www.kasa-amend.com/haselwander/index.html

AEG und Siemens, die hinter der Auseinandersetzung steckte, hatte den wirtschaftlichen Nutzen des Drehstroms längst erkannt. Haselwander blieb auf der Strecke. Eine schwere Enttäuschung, die der gebürtige Offenburger niemals überwand. Sein Freund, der Offenburger Drucker Franz Huber, sprach in Briefen von »Diebstahl« und bezeichnete Haselwander als »das betrogene Genie«...

Siemens bot Haselwander als Entschädigung fünf Mark pro PS an jedem verkauften Generator. Der Offenburger schlug das Angebot aus und übertrug das Patent an die Firma Lahmeyer—allerdings ohne Sicherung. Als die Firma aus finanziellen Gründen an AEG überging, ging Haselwander leer aus. [86]

Jonas Wenström (1889)[87]

The Swede Jonas Wenström (1855–1893) was born in Hällefors, Sweden. He suffered in his childhood, probably from the English disease, and his body became crooked and bent. When Jonas Wenström began his education at the college in Örebro, it was very clear that his talent was beyond the ordinary. In 1875 to 1879, he studied at the universities of Uppsala and Oslo. In Oslo he also began to correspond with Thomas Alva Edison. After completing his studies, he returned to Örebro to become a technical consultant at his father's construction firm.

He visited the International Electricity Exhibition in Paris and studied the modern dynamo machines presented. This stimulated him to develop his first electromagnetic machine in 1882 for which he obtained Swedish patent №. 409, granted on November 25, 1882, (also patented in

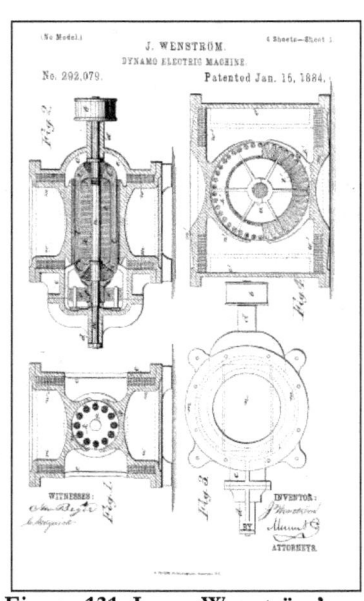

Figure 131: Jonas Wenström's US patent № 292.079 for a dynamo granted on January 15, 1884

Source: USPTO

---

[86] Source: http://www.schule-bw.de/unterricht/faecher/physik/ online_material/ e_lehre_1/stromsteckdose/drehstrom.htm ; and http://www.bo.de/lokales/offenburg/das-betrogene-genie.
[87] Source: Text based on autobiographic data in Swedish found at http://www.autor.se/ autoreter/ kraftverk/36.61.htm, http://runeberg.org/tektid/1928a/0467.html and http://www.abb.com/search.aspx?q=Wenstr%C3%B6m

other European countries and similar to US patent № 292.079 of January 15, 1884) (Figure 131). The machine had the shape of a turtle in order to create as little resistance as possible for the magnetic force field and, consequently, was called the "turtle."

Meanwhile, a businessman by the name of Ludvig Fredholm had initiated negotiations with the American Brush Company to buy the Swedish patent rights for a dynamo machine that the company made. Wenström offered Fredholm his patented design, which Fredholm bought for $20,000 and 10 percent of the sales revenues. On January 17, 1883, Wenström, together with Ludvig Fredholm, created the company *Electric Aktiebolaget* for the manufacture of electric-lighting DC installations with Wenström's dynamo. His brother George Wenström acted as the senior engineer and engineering manager. Jonas Wenström himself became technical adviser. The manufacturing itself was outsourced. On 2 March, 1884, the electric lighting system in Västerås Cathedral illuminated the interior for the first time. The Stockholm entrepreneur Ludvig Fredholm and the brothers Jonas and Georg Wenström lay behind this lighting system. Georg Wenström was in fact responsible, together with Fredholm, for installing, in 1881, Stockholm's first experimental street lighting system

Jonas was working constantly to improve his dynamo machines and presented a new dynamo in 1887. In April 17, 1888, he was granted US patent №. 381.451 (later followed by US patent №. 426.576 that was granted on April 29, 1890, and US patent № 515.386, granted on February 27, 1894, all for the design of a dynamo). The machine became a commercial success, and the company moved to new, larger premises. He also worked on a magnetic separator (for the separation of iron ore particles) that led to US patent №. 373.211, granted on November 15, 1887.

Table 18: Patents by Jonas Wenström for the AC induction motor, generator, and distribution system

Patent №	Granted	Description
US 292.079	January 15, 1884	Dynamo electric machine (filed December 7, 1882)—Equivalent to other European patents like Swedish patent 409 of November 25, 1882
US 373.211	November 15, 1887	Magnetic separator (filed on July 9, 1855)—Equivalent to Swedish patent 398 of December 24, 1884
US 381.451	April 17, 1888	Dynamo electric machine (filed on February 10, 1886)
US 426.576	April 29, 1890	Dynamo electric machine (filed on August 31, 1889)
US 455.808	July 14, 1891	Electromagnetic ore separator (filed December 24, 1890)
US 455.809	July 14, 1891	Magnetic separator (filed December 24, 1890)
US 515.386	February 27, 1894	Dynamo electric machine (filed on March 6, 1893)

Source: USPTO

Around 1889 Jonas Wenström started to develop his three-phase system, which comprised the whole-chain generator, transformer, and motor, as indicated in his laboratory-notes:

> *On October 6, 1889, three-phase generator calculated; on 14 November of the same year, the English patent description for the three-phase system; and on 20 January 1890, his Swedish petition filed under the heading: "devices for the circulation and dissemination of work through the use of three electrical alternating currents." The three-phase transformer and standard drive motor were pre-engineered in April and May 1890 [translated from Swedish]* ("Jonas Wenström ", 1928).

Jonas began experiments with an alternator[88] where his basic idea was to use not just one, but three alternating currents with a relative phase shift. Fredholm, then director of Electric AB, however, was only interested in lighting and did not want to devote time or money to the problem of the transmission of electricity. So George, together with mining engineer Gustaf Granström, created a new company in 1889: *Wenströms & Granströms Elektriska Kraftbolag*. Jonas, in the meantime, filed for several patents in 1890 related to a complete system for distributing electric power comprising a synchronous generator with star or delta connections, as well as transformers, synchronous motors, and induction motors, all of them designed to operate with three phases. In 1890 *Elektriska Aktiebolaget* merged with *Wenströms & Granströms Elektriska Kraftbolag* to form *Allmänna Svenska Elektriska Aktiebolaget*, later shortened to ASEA. In the same year ASEA, for which Wenström then served as a consultant, built and successfully tested the first prototype. Then, in 1891, Jonas Wenström visited the Electro-technical Exhibition in Frankfurt.

Figure 132: ASEA builds the first three-phase transmission system in Sweden: Hellsjön— Grängesberg (1894)

Source: http://history.vattenfall.com/the-revolution-of-electricity/before-vattenfall

> *There was an ongoing kind of competition between inventors in different countries about who would be the first to develop the three-phase system, and one of the competitors was Göran Wenström's brother Jonas Wenström. At a late stage, Jonas Wenström realized that he was about to lose the competition, and the proof*

---

[88] The word "alternator" was over time more and more used for the AC-dynamo generator.

of it was the Frankfurt-Lauffen system. The men behind it were Oskar von Miller and Michael Dolivo-Doborwolsky in Germany and Charles Brown in Switzerland. The patent was awarded in 1890, and Jonas Wenström's Swedish patent came in 1891. His patent gave ASEA numerous incomes and made the company a leading manufacturer in the field. However, the first tests of Jonas Wenström's three-phase machines in 1891 were not as successful as they had hoped. The motor was not constructed in an appropriate way. When Danielson returned the following year, he took on the problem and managed to solve it in a short time. This made possible the three-phase transmission of electricity between Hellsjön and Grängesberg in 1893, a mark in the history of Swedish electrification (Grönberg, 2003, pp. 124-125).

In 1893 ASAE built Sweden's first three-phase power transmission over a longer distance (15 km), between Hellsjön and Grängesberg (Figure 132). Jonas Wenström died December 18, 1893, at only thirty-eight years old. His legacy—the company ASEA—later merged into ABB, one of the giant European companies in the electric industry.

Brown's three-phase AC generator (1889)[89]

Charles Eugene Lancelot Brown (1863–1924), son of a Swiss mother and an English engineer, attended the Engineering School of Wintherthur and served an apprenticeship in the machine shop of *Bürgin & Alioth* in Basel, and later in his father's company (the *Schweizerischen Lokomotiv- und Maschinenfabrik*) before joining the *Maschinenfabrik Oerlikon* company in 1884. Two years later he became director of the electrical department at *Oerlikon*. In 1885 he conceived a type of armature winding for rotating machines now commonly used. The following year he undertook DC high-voltage transmission, developing several unusual features of equipment. Turning to AC machinery, in 1889 he designed some of the first oil-insulated transformers and produced generators suitable to the new requirements.

He was participating in the Lauffen project that supplied power to the Electro-technical Exhibition of 1891 in Frankfurt am Main. The Lauffen-Frankfurt project was essentially a joint venture of a German electrical company, *Allgemeine Elektrizitats Gesellschaft* (AEG), and a

Figure 133: The Oerlikon 220 kw generator used in Lauffen (1891)
Source: Deutsches Museum

---

[89] Text based on: *Charles Eugene Lancelot Brown*, Biography. Engineering and Technology Wiki
Source: http://ethw.org/Charles_Eugene_Lancelot_Brown

Swiss company, *Maschinenfabrik Oerlikon*. It was Brown, together with Walter Boveri, who built and was responsible for large projects at *Oerlikon*, including the 200 KVA generator for Lauffen project (Figure 133). Around that time Brown and Boveri decide to go into business together, and they create Brown, Boveri and Company (BBC)—a fast-growing Swiss company operating in a limited market, that would become one of the big conglomerates in the electric industry. In 1988 BBC merged with ASEA to create Asea, Brown, Boveri (ABB).

Later versions of induction motors

The properties that made the three-phase AC induction motor so attractive were its self-starting property and the constant torque. Soon after Dobrowolsky, others developed variations on the induction motor. Dobrowolsky himself continued working on the development and further improvement of the induction motor with a special rotor design: the "cage" induction motor.

> *Relying on his considerable knowledge of DC machines, Dobrowolsky built an experimental induction motor in the fall of 1888 with a "squirrel-cage" secondary... In 1889 he designed a 1/10-HP motor along these lines that had an efficiency of 80 percent and a relatively high starting torque*
> (Ronald Kline, 1987, p. 293).

Even the polyphase system saw its own versions of the induction motor when Stanley and Kelly developed the two-phase induction motor. They were granted US patent 505.859 for it on October 3, 1893.

> *Engineers later discovered through theoretical and experimental research that induction motors could perform well at high frequencies if they had correctly proportioned electric and magnetic circuits. William Stanley, Jr., and John F. Kelly, for example, invented in 1893 a widely used induction motor that ran from the same current as the Tesla-Scott motor: two-phase at 133 hertz*
> (Ronald Kline, 1987, p. 298).

The Westinghouse model C motor was a completely alternative induction motor designed by Benjamin Lamme in 1896. He used a technique of reducing the voltage applied to the motor when starting in order to limit the surge in line current.

> *The motor of 1895 was hardly settled as standard when a revolution came with the introduction of the type C motor. The characteristics of this motor were materially different from those of the motors in use, and it was much criticized, even inside the Westinghouse organization...The Westinghouse type C motor*

soon became the preferred type, and eventually took a prominent place in Europe. It advanced the induction-motor business enormously and created a reputation for reliability and durability of the induction motor, compared with the direct current, which placed the alternating motor far ahead of the direct current for general industrial purposes (Prout, 1921, pp. 126,127).

In the period after the invention of the induction motor, scientists started to more fully understand the fundamentals of the induction motor. Engineers in Europe and the United States published numerous papers and treatises on the mathematical theory of the motor. A few of the most noted were the following:

- Andre Blondel (1863–1938), born in Chaumont, Haut-Marne, France, was educated at the *École Nationale des Ponts et Chaussées* (School of Bridges and Roadways) in Paris. His studies and publications on the measurement of AC power, on the coupling of synchronous generators on a large AC electric grid, on the theory of synchronous generators, and other fields of electricity, made him well known. He studied the behavior of the arc light, created a method of calculation for induction motors, and designed the single-phase AC-DC commutator. He became a professor of electrotechnology in 1893 at the same school where he was educated: the *École Nationale des Ponts et Chaussées* in Paris (Capolino, 2004).
- Gisbert Johann Kapp (1852–1922), born in Vienna, Austria, studied mechanical engineering at the University of Zurich. After a visit to the Paris Electrical Exhibition in 1881, he worked in electrical engineering at the Crompton Works in Chelmsford (England). There he developed and patented the dynamo with compound field windings (GB Patent 4.810 of October 10, 1882). In 1881 he emigrated to England. And in 1886 he published on the theory of dynamo design and carried out design work on dynamos and electricity supply. In 1904 he became professor of electrical engineering at the University of Birmingham (Day, 2013).

In addition to the Europeans—such as those earlier mentioned: Dobrowolsky, Brown, Wenström, and others connected to the big European electric industries (i.e., Siemens & Halske, AEG, ASEA, BBC)—there were numerous others, both scientists and engineers, who contributed in the further development of the induction motor. In the United States, many of them were connected to Westinghouse and General Electric. Here are a few of them:

- Benjamin G. Lamme (1864–1924) was born in Springfield, Ohio, and he graduated with an engineering degree from Ohio State University. In

1889 he joined Westinghouse. He was the principal electrical engineer who built and improved generator designs from Tesla and was, arguably, Westinghouse's greatest pioneer. Lamme designed practical and reliable designs of all sorts of apparatus including generators, motors, and rotary converters (i.e., the single-reduction motor for street railways). He redesigned Tesla's induction motor, and today we use Lamme's induction motor design, not Tesla's. Lamme was responsible for designing the popular type-C induction motor, which went on the market in 1896, and was described at the time as having performance characteristics that were "unsurpassed" (J. C. Brittain, 1995).

Oliver Shallenberger (1860-1898) was born in Rochester, Pennsylvania, and he went to the Naval Academy at Annapolis at the age of seventeen. Among his contemporaries at the Naval Academy were Frank J. Sprague, Dr. Louis Duncan, W. F. C. Hasson, Gilbert Wilkes, and several others whose names are prominent among electricians. He became associated with George Westinghouse in 1884, worked on the Gaular & Gibbs transformer, and became Chief electrician at Westinghouse Electric Co. In fact, he was an early pioneer of AC power before Tesla and Lamme. At Westinghouse, Shallenberger was involved in the Great Barrington experiment with William Stanley. They pioneered transformer design. By 1895 Shallenberger was working with Westinghouse as an independent contractor ("consulting electrician"). His early contributions in the 1880s (including his AC meter in 1888: US patents № US 388.003, 388.004 and 426.335) were an important foundation for more sophisticated work in the 1890s (Terry, 1898).

Charles F. Scott (1864–1944) was born in Athens, Ohio. He graduated from Ohio State University and then worked at Westinghouse Electric & Manufacturing Co. in Pittsburg, where he was assisting Tesla and participated in the development of the Telluride, Colorado, project. He worked on high-tension transmission systems (50,000 volt) and developed the "Scott Connection" two-phase to three-phase transformation in 1894 (J. C. Brittain, 2002).

Charles Proteus Steinmetz (1865–1923) was born in Breslau, Silesia, and graduated from the University of Breslau where he was interested in socialism and joined a socialist group. Just before he was promoted at the university, he had to escape the German occupying forces who were after him. He emigrated to Switzerland in 1888. Like other political refugees from Bismarck, Steinmetz found a haven in the flourishing socialist life of Zurich. A regular in the salon of Carl Hauptmann, brother of Gerhardt, the noted playwright, Steinmetz retained his interest in socialism while studying engineering at the

Zurich Polytechnic. After his permit expired, he emigrated to the United States in 1889. When his employer at that time (Rudolf Eickmeyer, also a German immigrant) merged with General Electric, he became GE's engineering wizard. His contribution was in the field of AC systems theory, and he became well known for his publication of magnetic hysteresis. Steinmetz obtained more than two hundred patents in his lifetime (R. Kline, 1987).

All these contributions, both to the theoretical basics and the practical engineering, stimulated the rapid development of the induction motor—followed by its use in a wide range of commercial and industrial applications.

> *By the mid-1890s, the theory of the induction motor had caught up with practice. No longer were engineers dependent on the type of cut-and-try procedures followed by Tesla, Scott, and Dobrowolsky. Through the work of Kapp, Steinmetz, and Behrend, they now had two methods—the equivalent circuit and the circle diagram—that placed the design of induction motors on a more rational basis…The early fruits of this theory were impressive. From the mid-1890s onward, American and European engineers designed an enormous variety of induction motors for a wide range of applications, from turning desktop fans to propelling battleships* (Ronald Kline, 1987, p. 310).

The invention of the AC induction motor

A range of scientists and engineers contributed to the development of the AC induction motor: from Ferrantis to Haselwander, from Tesla to Bradley, Dobrowolsky, Brown, and others. Many of these claimed to be the inventors—claims that resulted in a lot of disputes. It is not within the scope of this case study to evaluate these disputes, but, for the purpose of our analysis, we can conclude the following.

Taking a helicopter view of the matter, it all depends on the definition of the invention itself and the criteria to be used in the judgment. Are we talking about the components of the system (i.e., the induction motor) or the total system itself (i.e., the principle of polyphase AC systems)? Are we considering the principle itself, the prototype proving that is was feasible, or are we considering the finished product introduced (successfully) to the market? Do the rights of invention occur at the moment the inventor published about his work, at the moment that he demonstrated it publically, or at the moment he got his patent? Or was it not the patent that mattered at all, but the economic, technical impact of his discovery that made him the real inventor?

From a *legal point of view*, looking at the patents issued, Haselwander failed with his patent application in 1885, but he did build a three-phase AC generator and motor. Ferrantis did not file for a patent but published about the induction motor in 1888. Tesla did file several patents on polyphase systems and had been building a two-phase system; however, it was a system that failed in the market. That being said, one can note that in literature Tesla is considered to be a major influence in the development of the induction motor. Also, there were several court cases where Tesla's priority was at issue: *Westinghouse Electric Mfg. Co. vs. New England Granite Co.* (1901), *Westinghouse Co. v. The Catskill Illuminating Co.* (1899), and *Westinghouse vs. Dayton Fan and Motor Company* (1905). This legal activity indicates that Tesla's patents were considered worth attacking. In the lawsuit *Westinghouse Electric & Mfg. Co. vs. Mutual Life Ins. Co.* (1904), the final verdict did, in the end, favor Tesla. The judge ruled as follows:

> *The patents in suit especially have been attacked with well-directed, vigorous, and resolute pertinacity. The fundamental principles upon which a difference of phase in circuits is based have been set forth with elaborate detail in prior opinions by Circuit Courts and Circuit Courts of Appeals, notably by Judge Townsend in the case of Westinghouse v. New England Granite Co. et al. (C. C.) 103 Fed. 951, which was a suit upon the broad Tesla patents of May,1888, nos. 381,968, 382,279, and 382,280...*
>
> *The Circuit Court, considering the Tesla patents in suit and the defenses there raised, sustained their validity, and unqualifiedly concurred in the decisions of Tesla Electric Co. v. Scott & Janney et al. and Westinghouse Co. v. Dayton Fan & Motor Co., supra. The Circuit Court of Appeals, however, reversed the decision upon the ground that the publication of a magazine article on April 22, 1888, by Prof. Galileo Ferraris, fully described and disclosed the System covered by the patents in suit. This publication upon the evidence in that case was found to be prior to the date of the inventions in suit, and constituted an anticipation...Upon careful consideration of the proofs, I have arrived at the conclusion that the actual date of the Tesla inventions is prior to this publication, and that the patents were not void for anticipation. According to the evidence, Tesla conceived his invention in his laboratory. №. 89 Liberty street, New York City, and completed the same in the month of September 1887...*
>
> *The standard of proof required, where anticipation has been clearly shown, to carry the invention back to a date earlier than the application, has been abundantly supplied in the present record. Here the testimony of Tesla, emphatically and unequivocally narrated, sufficiently supported by other witnesses,*

as to the specific construction of the exhibit motor and its operativeness as a split-phase derivative motor in the month of September 1887, impels me to the conclusion that its actual invention is prior to the date of the Ferraris publication…These are all significant facts, which in my judgment supply the definiteness and certainty on the question of priority of invention, which the court found absent in the Catskill Case. For these reasons, the date of the inventions in suit is carried back to September 1887. ("Cases argued and determined in the Circuit Courts of Appeals and Circuit and District Courts of the United States. ," 1904, pp. 215-219).

Looking from a *technical point of view*, Ferraris's work was essential in proving the concept. And the development of the principle of the induction motor into a working device was certainly stimulated by Tesla's efforts in 1887. His work, which was also based on the work of others, was quite fundamental for the actual creation of induction motors.

The century-old priority dispute about who invented the motor is unresolved and will likely remain so. But it is clear that two men—Nikola Tesla and Galileo Ferraris—went far beyond other claimants. Tesla made the first successful patent application (November 1887), and Ferraris first announced the principles of the motor (March 1888). Both men, however, were familiar with only part of the previous work on the science of their device. They knew about Arago's disk because it had been a common electrical experiment since the 1850s. But there is no evidence that they, or other inventors, drew on the research of Baily and Deprez, which was not well known until the early 1890s. Instead, Ferraris and Tesla independently worked out how to produce a revolving field electrically and combined this knowledge with that of Arago's rotations to invent the induction motor…Hence, Tesla's major improvement over Ferraris—the use of closed-circuited windings—resulted, ironically, from his misunderstanding of Arago's rotations (Ronald Kline, 1987, pp. 288, 291).

In terms of realizing a successful product, it was Dobrowolsky and Brown who realized the impressive working systems in 1888 and demonstrated them successfully at the Frankfurt Electrical Exhibition in 1891.

Tesla was the first to work intensively on electric power transmission through a multi-phase alternating current system, he was the first to find the basics for such a transfer and was the first to present the principles of a multi-phase induction motor. Bradley filed the first patent on a two-phase AC power transmission system with synchronous machines and four electric wires. He also created the first patent for a three-phase induction motor with a completely shorted-circuited rotor winding (squirrel-cage induction). Haselwander was the first to design a three-phase transmission system with three-phase synchronous machines and three transmission lines. He built the first such facility, and gave it first into practical use. Dolivo-Dobrowolsky built the first simple, practically useful three-phase induction motor with squirrel cage rotor. In broad scientific lectures and essays, he explained nature and characteristics of the three-phase current system and three-phase motors...and two years after the construction of his first 1/10 hp induction motor, he put a 100-horsepower three-phase motor into normal operation...Dolivo-Dobrowolsky must therefore be seen as the pioneer for the introduction of the three-phase current system.[90]

So, one certainly can conclude that the invention of the induction motor, seen from different points of view, cannot be attributed to one inventor. Considering the whole system (generation, transmission,

Table 19: Patents related to the AC induction motor (pre-1889/1890)

Patent №	Year	Patentee	Description
US 381.968	May 1, 1888	N. Tesla	Electromagnetic motor: polyphase AC induction motor (filed October 12, 1887) Assignor: Charles F. Peck
US 382.279	May 1, 1888	N. Tesla	Electromagnetic motor: polyphase AC induction motor (filed November 30, 1887) Assignor: Charles F. Peck
US 388.003	August 14, 1888	O. B. Shallenberger	Meter for alternating currents (filed on June 6, 1888)
US 404.465	June 4, 1889	C. Bradley	Electric motor (filed on October 5, 1888)
US 413.986	October 29, 1889	C. J. van de Poele	Alternating current induction motor (filed on July 29, 1889)
US 426.335	April 22, 1890	O. B. Shallenberger	Armature for electric meters (filed January 16, 1890)
US 427.978	May 13, 1890	M. Dolivo-Dobrowolsky	Alternating current motor: method of producing rotative motion by means of alternating electric currents proposed by Professor Ferraris in Turin (filed November 13, 1889)*
GE 51.083	March 8, 1889	M. Dolivo-Dobrowolsky	Idem US 427.978*
FR 199.154	June 24, 1889	M. Dolivo-Dobrowolsky	Idem US 427.978*

* Equivalent to other European patents like German patent 51.083 dd. March 8, 1889; French patent 199.154 dd. June 24, 1889

Source: USPTO.

distribution, and usage of AC power), the development of the AC system (generator and motor) had many participants—most of whom replicated another's work, sometimes simultaneously, often with no knowledge of the other. Looking at the induction motor itself, it was certainly the efforts of Dobrowolsky and Tesla that made the induction motor a working machine.

Patent activity

All the described activities, experiments, and developments for the polyphase and three-phase AC induction motor have resulted in a range of patents, indicating much innovative activity. In Table 19 some of the US patents are shown that can be considered as being more or less important to the development of the AC induction motor up to Dobrowolsky's US patent № 427.978, a patent that also covered the engine in other countries. In Table 20 the same information is presented for the patents that are identifiable after Dobrowolsky's patent.

Table 20: Patents related to the AC induction motor (1890–1894)

Patent №	Year	Patentee	Description
US 456.804	July 28, 1891	M. Dolivo-Dobrowolsky	Alternating current motor (filed on December 23, 1890)
US 471.155	March 22, 1892	E. Thomson	Alternating current motor (filed on August 17, 1891)
US 514.904	February 20, 1894	Ch. Bradley	Alternating current motor (filed on January 14, 1893)
US 518.310	April 17, 1894	T. Duncan	Universal phase alternate current motor (filed on May 22, 1893)
US 529.272	November 13, 1894	M. Huttin, M. Leblanc	Alternating current electrodynamic machine (filed on August 20, 1892)

Source: USPTO

A cluster of innovations for the AC induction motor

The single-phase AC motor was not too successful in its infancy due to its limited performance. The same goes for the double-phase induction motor, which needed a complicated infrastructure that was different from the status quo. The DC motor, however, after years of struggling in its development, got a revival that was heralded by the work of Sprague. These motors were especially successful in motive applications (i.e., streetcars) with their own power-supply system.

The breakthrough for the AC motor came when three-phase AC power could be economically transported over large distanced without too much loss. The resulting three-phase induction motor—as conceptualized and shaped by Tesla and Dobrowolsky—in combination with the powerful distribution system, became the dominant design for a range of AC induction electric motor designs. The key factor was the infrastructure of the power-distribution system.

Soon a range of improvements followed the development of the three-phase AC induction motor concept, among which was the squirrel-cage induction motor and the Westinghouse Type C motor (Figure 134).

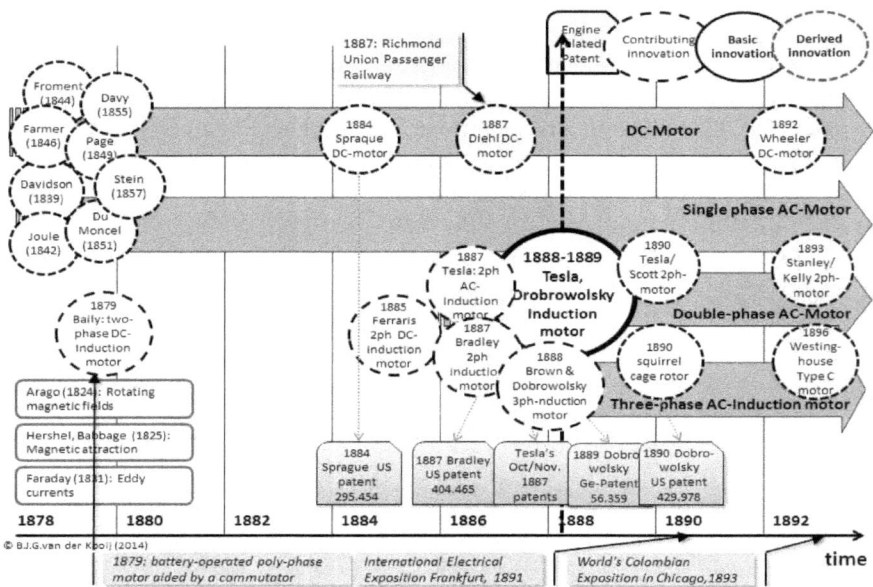

Figure 134: Cluster of Innovation around the Tesla, Dobrowolsky induction motor
Source: Figure created by author

The development of the electric alternating-current power system

The development of the induction motor was a breakthrough for the further development of electricity as a power source. Its development went hand-in-hand with the development of the three-phase AC generator. Together with other components such as the transformer, they created another breakthrough: the transmission of electricity over longer distances. The European developments (culminating in the Lauffen project and shown at the Electrical Exhibition in Frankfurt am Main in 1891) proved that the AC power system, both from the point of electricity generation and the use of the induction electric motors, was the way to go.[91] It was the result of the development of the three-phase electric power system (in the years 1880 until 1890) that became the basis for modern electrical power transmission and advanced electric motors.[92] This development would create fast networks for the distribution of electric power over the following decades (Hughes, 1993).

The components of the alternating-current power system for electricity have been presented in this study: the AC generator, the transformer, and the all-important induction motor. After the development of these important components was explored, the early development of the

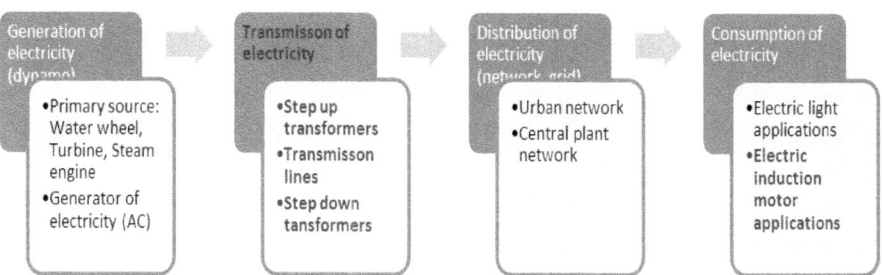

Figure 135: **Overview of the AC electric systems of generating, distributing, and consumption of electricity**

---

[91] As an induction motor's operation depended on the interaction between an electromagnetic field produced by currents fed to the motor's "primary" coils (or windings) from an outside source—the electric dynamo—and an electromagnetic field produced by currents induced in "secondary" coils or conductors by the rotation ("slip") of the primary field relative thereto, this outside source was to be the three-phase AC generator.

[92] As these motors rotated on the rotating fields created by the three-phase generators (which were also rotating, powered by their primary source), they were called *synchronous induction motors*.

electricity distribution system was described. From local plant to central station, the development of electrical systems started with the small, local, and urban, mostly DC-based distribution systems. These were almost exclusively used for lighting systems and for the occasional DC-motor application (i.e., the elevator). After that, we traced the beginnings of the small mixed AC-DC systems and the (single-phase) AC systems.

> *When the alternating current was introduced for practical purposes, it was not needed for arc lighting, the circuit for which, from a single dynamo, would often be twenty or thirty miles in length, its current having a pressure of not less than five or six thousand volts. For some years it was not found feasible to operate motors on alternating-current circuits, and that reason was often urged against it seriously. It could not be used for electroplating or deposition, nor could it charge storage batteries, all of which are easily within the ability of the direct current. But when it came to be a question of lighting a scattered suburb, a group of dwellings on the outskirts, a remote country residence, or a farm house, the alternating current, in all elements save its danger, was and is ideal. Its thin wires can be carried cheaply over vast areas, and at each local point of consumption the transformer of size exactly proportioned to its local task takes the high-voltage transmission current and lowers its potential at a ratio of 20 or 40 to 1, for use in distribution and consumption circuits* (Dyer & Martin, 1910, p. 421).

AC-distribution networks

Slowly the AC power systems appeared on stage, gaining more popularity due to their ability to transmit over larger distances economically. However, it took a while before the "universal" high-voltage AC system existed. This system developed in phases: from the single-phase AC system, through the polyphase systems, to the three-phase AC system we use today.

As explained before it is not only the development of the components (such as the dynamo generator and the electric motor) that is relevant to electric power systems but also the development of the system as a whole. And alternating current added an additional "subsystem" to the generation

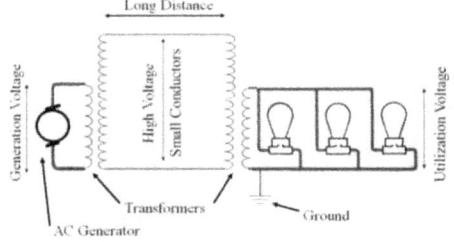

Figure 136: Adding transmission in AC systems

Source: http://www.ieeeghn.org/wiki/ images/5/53/02-Edison_Central_Station_3_ wire_dc_system-21.GIF

system and the distribution system: the *transmission system*. In this system electricity was brought to a higher voltage with a step-up transformer, transmitted over a long distance, and then brought down to the utilization voltage with a step-down transformer. So the electric system became extended by a fundamental component: the transmission of AC electricity (Figure 136).

AC systems: patents

The second half of the 1880s and the first half of the 1890s were the years in which the AC system was developed. It was George Westinghouse who adopted alternating current and made it the cornerstone of his entrepreneurial activities. Employing people such as Nikola Tesla, Oliver

Table 21: Patents granted for AC systems of electrical distribution controlled by George Westinghouse

Patent №	Granted	Description
US 373.035	November 8, 1887	System of electric distribution: AC system with storage battery for storage of current to be used against an emergency
US 381.970	May 1, 1888	System of electrical distribution: Four-wire, two-phase AC system with independent rotating fields, independent electric transmission circuits, and transformers with primary and secondary coils (patented by N. Tesla, filed December 23, 1887)
US 382.282	May 1, 1888	Method of converting and distributing electric currents: two-phase AC system with independent rotating fields (patented by N. Tesla, filed December 23, 1887)
US 390.413	October 2, 1888	System of electrical distribution: Two-phase AC system with independent rotating fields, with one conductor as return wire in common—three-wire system (patented by N. Tesla, filed on April 10, 1888)
US 390.990	October 9, 1888	System of electric distribution: AC system connecting a number of translating devices in series with each other, and in connecting in shunt-circuit around each device a suitably constructed reactive coil, and in thus dispensing with the converters or potential-reducing devices as usually organized (patented by Oliver B. Shallenberger, filed on October 1, 1887)
US 428.651	May 27, 1890	System of electric distribution: system of distribution by alternating currents with compensating coils (patented by E.Thomson on August 15th, 1888)
US 468.122	February 2, 1892	System of electric distribution: electric distribution in which lamps or other translating devices are arranged in the well-known multiple-series fashion; and its purpose is to provide an effective means of maintaining a fairly equal potential on the different branches of the system, notwithstanding inequalities of load which may exist therein. (patented by Elihu Thomson: filed on March 11, 1891)

To be continued on next page

Shallenberger, and Edwin Rice, he controlled important patents for the development of the electrical AC system.

Nikola Tesla was granted patents for his polyphase system, for example US patent № 381.970, granted on May 1, 1888, for a "two-phase AC system with independent rotating fields." And Houston-Thomson Electric, with Elihu Thomson and William Stanley, developed systems and obtained patents. Besides the patents that Westinghouse controlled, others were developing parts of the power system—for example the components for

Table continued from preceding page

US 503.621	August 22, 1893	System of electric distribution: The general plan of the invention is to generate currents of such potential as may be economically produced, and to then increase the potential to such a degree as to render it possible to transmit a large amount of electrical energy over conductors of small cross section, and to subsequently reduce the potential of these currents, at or near the points where they are to be consumed, to such a degree as may be desired, for operating the special translating devices. (patented by Oliver B. Shallenberger, filed on March 3, 1887)
US 503.622	August 22, 1893	System of electrical conversion and distribution: Wherein alternating, and pulsatory, intermittent, or pulsatory electric currents of any required potential, and derived from any convenient source, are transformed or converted, in whole or in part, into secondary currents having a different potential. (patented by William Stanley, filed on March 28, 1887)
US 508.839	November 14, 1893	System of electric distribution: Three-phase system using a step-up transformer and a step-down transformer at the point of distribution (patented by Edwin Rice: filed on May 31, 1893)
US 519.076	May 1, 1894	System of electric distribution: To secure on the local circuit freedom from danger of shock to persons as well as immunity from fire on such circuits due to the accidental existence therein of the high potential current passing upon the main or feeding circuit and which high potential current may accidentally find its way to the secondary circuit by a leak or abnormal connection. (patented by Elihu Thomson: filed on August 5, 1889)
US 521.051	June 5, 1894	System of electric distribution: transmission of electric currents by the number of phases best securing the highest economy and convenience and involves the change to this number of phases from the number generated at the source of current, or the change from the number used in transmission to the number desired for translating devices, or both changes. (patented by Charles Scott: filed on February 26, 1894)

Source: USPTO

the distribution system as described in US patent №. 380.757 granted to W. H. Hart and J. Th. Goodfellow on April 10, 1888, for a system of underground cabling. All these efforts resulted in projects that implemented the concepts and equipment. Here are some examples of early AC-transmission projects.

Early AC-transmission projects

Single-phase AC dynamos were originally used to generate electricity in remote, rural areas from waterpower and to transport it over longer distances to factories or urban areas—such as the Ames Hydroelectric Generating Plant in 1891 (Figure 137). This plant was built because the gold mining industry at Telluride, Colorado, was to be shut down due to the depletion of cheap steam power. All timber in the area had been cut for fuel and for mining timbers. DC electric power and other forms of power transmission had proven to be ineffective to meet the 4.2-kilometer distance of needed transmission. AC power was judged to be the only workable solution to the economic problems of the mining industry.

Figure 137: Powerhouse of the Ames Hydroelectric Generating Plant
Source: US. Library of Congress
www.loc.gov/pictures/item/co0030.photos.021928p/

Again, the problem was manifest, and the solution had become available. As financing was not a problem, it became, at 3,000 volts and at 133 Hz, the first high-voltage application in the United States ("Ames Hydroelectric Generating Plant," 1891).

> *One of the first industrial high-voltage AC power systems was erected in the mining district of Telluride, Colorado, with power transmission from two waterfalls over a distance of about 4 km (2.5 mi) to the Gold King Ore Mill, 610 m (2,000 ft) higher up* (Neidhofer, 2007, p. 89).

This example of an early project shows the characteristics of the "isolated plant" system for incandescent lighting, with the only difference that the transmission was placed between the generation and the distribution of electricity. In the same evolution, we see the "general station concept" adapting AC electricity, such as the central plant in Buffalo:

> *The first alternating current central station to operate commercially in the United States was placed in service in Buffalo on November 30, 1886, only four years after Edison's Pearl Street Station. It was a Westinghouse 400 lamp single-*

Figure 138: View of power development by the Niagara Falls Hydraulic Power and Manufacturing Company
Source: http://library.buffalo.edu/pan-am/exposition/electricity/development/

phase system (also called two-wire system) with a primary of 1,000 volts. The generator was located in the Brush Electric Light plant at Wilkeson and Mohawk Streets. One customer was the Adam, Meldrum & Anderson department store on downtown Main Street, now the site of the Main Place Mall ("Early Electrification of Buffalo," 2004).

All these projects had either a steam machine or a waterwheel as primary movers. The latter, water driving turbines, would become an important source when the Niagara was utilized for electricity generation.[93] The Niagara Water Falls (Figure 138), a tremendous source of hydropower, was already used in 1879 for Brush arc-light generators (DC). In 1893 the Niagara Falls Power Company decided to adopt two-phase AC and ordered three hydro generator units with remarkable ratings of 5,000 hp each. Late in August 1895, the power

Figure 139: The Niagara Falls Power Company's power line between Buffalo and Niagara
Source: IEEE Global History Network.
http://www.ieeeghn.org/wiki/index.php/File:0 7-93_new_wood_pole_line.GIF

---

[93] For more details see: http://library.buffalo.edu/pan-am/exposition/electricity/development/

system (operated by the Cataract Power and Conduit Company) went into operation. The system provided electricity to cities such as Buffalo, in the state of New York, via power-transmission lines (Figure 139).

It was the zenith of the two-phase system as it soon became obsolete. From 1895 the Niagara Falls would be used to power three-phase generators). The electricity was used to power the industries in Buffalo, twenty-two miles away, and replaced the former local power plants. In 1897 the steam plants of the various electric power companies (in Buffalo) were gradually dismantled, and the power was taken from Niagara Falls through The Cataract Power and Conduit Company.

In Germany one of the early AC high-voltage projects was the Lauffen am Neckar-Frankfurt am Main project used to power the 1891 International Electro-Technical Exhibition in Frankfurt (Figure 140). This was a three-phase AC system that transported electricity over 175 km. In 1894 the German company AEG built power stations based on three-phase techniques in increasing numbers for the home market and for export. Siemens & Halske and General Electric began work on three-phase generators the same year as the exposition (Figure 141).

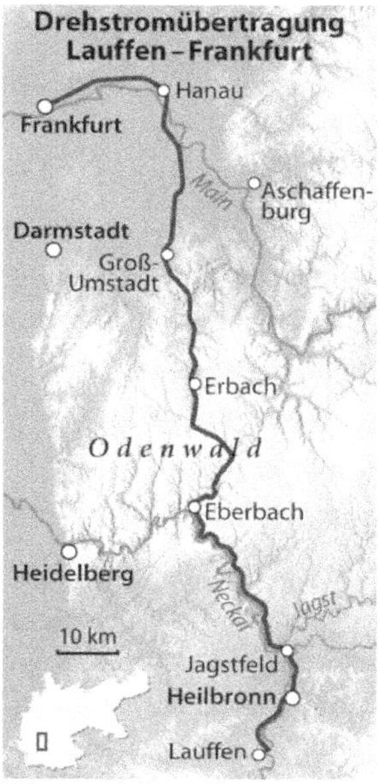

Figure 140: AC transmission between Lauffen and Frankfurt (1891)

Source: Wikimedia Commons

Figure 141: The generator manufactured by Oerlikon in Lauffen am Neckar (1891)

Source: Wikimedia Commons

These are just a few of the early projects in which alternating current was used for the transmission of electricity with high voltage over longer distances. But in the end, three-phase AC electricity would become the standard of electrical power.

The adoption of the new technique was also accompanied with problems of patent rights and partially handicapped by litigation for years. Thus, the breakthrough had its own particular course in the countries concerned...The battle of the currents, in its last period, was a showdown of various classes of AC, from single-phase AC up to polyphase variants. Depending on the circumstances in the leading countries, initial progress was first made by two-phase systems or directly by the three-phase system, which definitively won the competition. From approximately 1900, it was clear that three-phase power was going to become the leader in electricity supply throughout the world (Neidhofer, 2007, p. 100).

George Westinghouse (1846–1914)

George Westinghouse (1846–1914), born in Central Bridge, New York, was the son of a machine-shop owner. He was descended from Westphalian stock (Germans with the family name Wistinghausens) that came to America in 1755. During the Civil War (1861–1865), he enlisted in the army at the age of sixteen, and he later joined the navy, after which he went to Union College in 1865. He was not a student for long, however, and he dropped out of school three months later and went to work in his father's workshop in the city of Schenectady. The shop's business was the making of agricultural machinery, mill machinery, and small steam engines (Prout, 1921, pp. 6-8).

Westinghouse's early business activities

While in Schenectady, New York State, he worked on steam engines and obtained his first patent (US patent № 50.579 on October 31, 1865). Patents related to railroad equipment followed this patent. He also worked on air brakes for railroad cars and developed a system using compressed air in 1868, when he was twenty-two years old. Air brakes were quite different than the mechanical systems that were commonly used to stop trains at that time.[94] His first

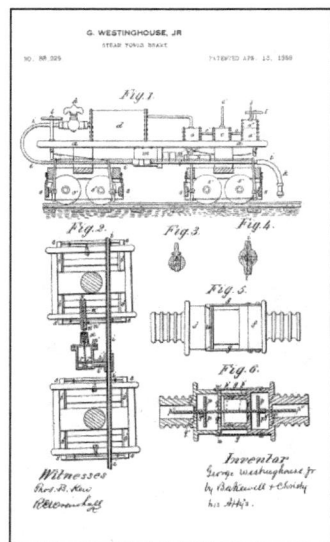

Figure 142: Westinghouse airbrake patent № 88.929 (1869)
Source: USPTO

---

[94] Prior to his invention, engineers controlled steam to propel the train, and brakemen rode atop the cars in the open air, scrunching down as they passed through tunnels, jumping from car to car to turn a wheel on each one that tightened a set of chains and applied the mechanical brakes, one car at a time, bumping and grinding their collective way to an

air-brake patent, US Patent №. 88.929 granted on April 13, 1869,[95] was filed from Schenectady (Figure 142).

Between 1868 and 1869, Westinghouse completed the original development of an air-brake system (covered by some twenty patents) and started the Westinghouse Air Brake Company, capitalized[96] at $500,000 in 1869. In early 1870 his company was already struggling to meet the orders. But it was a major train crash that resulted in Westinghouse's breakthrough in the US market. On February 6, 1871, the Pacific Express ("America's Number One Train" as it was called) collided near Wapinger Creek Bridge (near Poughkeepsie, between Chicago and New York) with a southbound freight train. The Pacific Express ploughed into the oil-tank cars, resulting in massive explosions and the deaths of twenty-two people. The result was a stampede of railroad companies wanting air brakes on their passenger trains. Westinghouse continued to improve the brake system, designing a fail-safe air-brake system. He applied for patents, resulting in a range of US patents. for example №. 117.841 on August 8, 1871; №. 122.404 and 122.405 on March 5, 1872, and №. 124.405 on March 5, 1872. The implications of the automatic air brake were staggering—not only in terms of safety, but also in economic terms. Longer trains running at higher speeds caused rail-haulage costs to plummet.

During the 1870s George Westinghouse spent the greater part of his time in Europe. One of his aims was to sell his air-brake system to British railway companies. Always improving and adapting the air-brake system, in 1872 the Westinghouse Continuous Brake Company was established in New York City to manage the European export business, eventually resulting in the formation of the Westinghouse Brake Company Limited in England in 1881.

> *After his first trip to Europe, he returned and introduced a five-and-a-half-day week. He established this in a time when they were having difficulties fulfilling orders. Half a Saturday off for family and Sunday for God was unheard of in an industrial enterprise of the time...Westinghouse did these things from his heart,*

---

operationally precarious and acoustically rackety halt. Source: http://www.pghtech.org/news-and-publications/teq/article.aspx?Article=1759

[95] This patent was contested in 1874, and Westinghouse was accused of not being the original inventor. On June 16, 1875, the court decided in favor of Westinghouse (Prout, 1921, p. Appendix Patents).

[96] When a new company is incorporated its total number of shares has a nominal value. For example 500.000 shares at $1,- each. This is meant by "capitalized". The shares sold create the working capital, that can be considerable lower; e.g. $100.000 when only 100.000 shares are issued. However, when the stock is traded at a stock exchange, it has a specific market value: e.g the share price of $8, 27. This gives the "market capitalization": a company's outstanding shares multiplied by its share price. .

> *but many claimed the motive was to increase productivity. To Westinghouse, the increase in productivity was a side benefit of doing the right thing. In any case, productivity did boom through good management, good engineering, and a caring approach to his employees. Westinghouse never changed his deep concern for the employees. Westinghouse never saw himself as other than a fitted representative of God, while other great industrialists, such as Carnegie, viewed themselves as trustees of God...He sincerely cared about his fellow man*

(Skrabec, 2007, p. 54).

His first enterprise, Westinghouse Air Brake Co. (1869), was to be followed by an avalanche of companies: Westinghouse Machine Co.; Westinghouse Foundry Co.; Union Switch Co.; Philadelphia Co.; Westinghouse Electric Co.; Westinghouse Electric & Manufacturing Company. His success in applying air brakes on trains all over the world was massive. In 1876 Westinghouse was a world-class, international business.

> *As soon as the air brake was fairly underway in America, Westinghouse took it to England, and within ten years, that is, before he was thirty-five, he had organized and established shops in England, France, and Russia. He was famous and had a fortune sufficient for his moderate needs*

(Prout, 1921, p. 15).

Even though Westinghouse was becoming quite a business entrepreneur, he still loved working on the steam engine. From 1870 to 1875, he continued to work on rotary steam engines. He improved and patented the steam "governor" (a regulator for the steam engine, US patent № 162.782 granted on May 4, 1875). But he did more.

> *Like Edison, Westinghouse enjoyed working on many projects simultaneously. In the mid-1880s, not only was Westinghouse developing the modern-day system of natural gas to replace coal smoke, an urgent concern in his beloved but sooty Pittsburgh, he was also working out a system of pneumatic transmission of power, early experimenting on refrigeration, and...beginning experiments with electric light and power* (McPartland, 2006, p. 287).

In 1884 Westinghouse, as a logical extension of his fail-safe air-brake system, went on to address the next pressing issue of the burgeoning railroads: how to prevent collisions between trains traveling along the same stretch of track, either head-on or engine-to-caboose. He did this by devising a signal system that employed the tracks themselves as an electrical conductor: the block signaling system. He invented the electropneumatic device for switch and signal, acquiring patents for it. From 1881 to 1882,

Westinghouse patented eighteen inventions in switch and signal control (among those US patent №. 240.628, US patents №. 245.108, №. 245.592, and №. 246.053). To industrialize these inventions, he bought two companies—the Union Electric Signal Company and the Interlocking Switch and Signal Company, both having essential patents—and combined them into the Union Switch & Signal Company in May 1881.

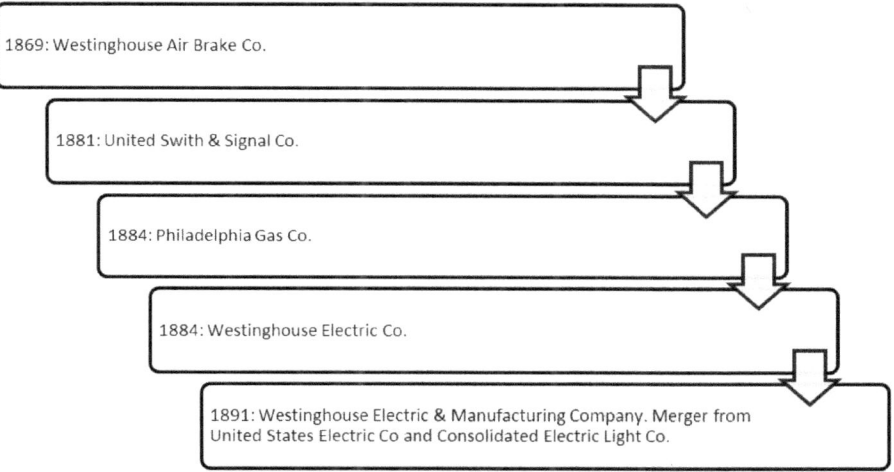

Figure 143: Successive companies created by George Westinghouse

Late in 1883 Westinghouse became interested in the production and distribution of natural gas. Natural gas was available in abundance in Pittsburg. He started drilling in the backyard of his house "Solitude," and in May 1884 he hit a gas well.

> *The gas rose into the air with such force that it tossed aside heavy objects thrown into it by spectators, and large pieces of coal or even planks were splintered by the pressure. A hundred-pound stone was lowered from the derrick by a rope, but it was thrown to one side. The day was given up to finding methods of restoring the grounds to their former state, but this was difficult because the drillers were not sure how the new well would behave. Although they had drilled many wells, this was the most startling performance they had witnessed...Encouraged by his spectacular success, Westinghouse drilled several other wells in the surrounding district and the Solitude bore came to be known as Westinghouse Well №. 1 to distinguish it from the others. A craze for drilling wells seized the city; as one Pittsburgh newspaper, speaking of the East End district alone, put it laconically, "Twenty gas wells on the tapis"* (Van Trump, 1959, p. 165).

Westinghouse created the Philadelphia Company in 1884 to supply gas to the residents of Pittsburgh. The inventive Westinghouse acquired thirty-eight patents covering gas distribution and regulation devises. Among those are US patents №. 301.191 for a "System for conveying and utilizing gas under pressure" granted on July 1, 1884; US patent №. 314.089 for a "System for the protection of railroad-tracks and gas-pipe lines" granted on March 17, 1885; and US patent №. 315.363 covering "Means for detecting leaks in Gas-mains," granted on April 7, 1885. All his work gave him the experience to build experience with distribution systems. This was an advantage as, not too much later, the gas-lighting industry would be facing its new competitor, electricity, with its electric candles.

> *The introduction on a large scale of natural gas into Pittsburgh manufacturing had enormous consequences for the industrial future of the city...New steel and iron industries were attracted to this area that might well have gone elsewhere, and it was Westinghouse who helped materially to make Pittsburgh one of the great industrial cities of the world* (Van Trump, 1959, p. 167).

During the first two decades of his career, George Westinghouse was active in his father's machine shop. Then he developed a business on his own. He engaged mainly in the fields of air brakes, railway switches, and signal systems and became a leading figure in those fields. These experiences helped him build considerable technical and entrepreneurial experience. The number of US utility patents that he acquired in his lifetime totaled 353, not counting reissued patents. Of these, 131 patents—applied for prior to 1885—were classified in air-brake-related groups, switch-and-signal-related groups, and in the natural-gas group (Nishimura, 2012, p. 7).

Westinghouse enters the electricity business

So, in the early 1880s, Westinghouse was active in air-brake systems, railway-signaling systems, and natural-gas distribution. It was the time when the arc light, powered by DC generators, was gaining popularity, and this drew his attention and was a development that did not escape Westinghouse's entrepreneurial consideration.

> *Late in 1883 Westinghouse began to think somewhat seriously about direct-current lighting. He began to gather about him a staff, and soon had several men busy in study of methods and in development of details; but not until he had his vision of the possibilities of the alternating current was his interest thoroughly aroused* (Prout, 1921, p. 91).

Alternating current of high tension could be used for distributing electricity economically along greater distances (and could use much less of the expensive, thick copper wires that the DC systems needed), but it needed additional facilities to make it practicable—for example equipment to bring the voltage down to a safer voltage that would be usable in home and office environments. So Westinghouse started by paying attention to the issue of transforming electrical currents. In 1884 he contracted with Stanley to join the staff of the Union Switch and Signal Company and to engage in the development of incandescent lighting systems. Stanley, with E. P. Thomson, had earlier worked on an incandescent lamp with a filament of carbonized silk. He had also worked on a self-regulating direct-current dynamo and had several patents for induction coils (today called electric step-up and step-down transformers)—such as US-Patent №. 349.616 granted on September 21, 1886.

> *In 1885 he [Westinghouse] became interested in electrical transmission using an alternating current. Learning of a type of transformer developed in Europe by Gaulard and Gibbs, he promptly ordered the transformer and Siemens's alternator. As early as November 1885, the transformer and generator were transferred from London with Reginald Belfield, who was an assistant of Gaulard and Gibbs. Westinghouse began to develop a practical transformer by examining the imported transformer with Stanley, Belfield who was employed by Westinghouse, and other staff…In February 1886, when a prospectus of the Electric Company had already been prepared, Westinghouse dispatched Pope and Guido Panteleoni, who was employed by him, to England to secure the American rights to the inventions of Gaulard and Gibbs. They successfully negotiated with the inventors and contracted for $5,000. The petition for the patent was filed on March 6th 1886; the patent, entitled "System of electric distribution," was issued on October 26th of the same year. His self-regulating transformer would become a cornerstone in the development of the distribution system for AC-electricity. In March 1886 the electrical department of the Union Switch and Signal Company broke away to become the Westinghouse Electric Company, capitalised at $1 million* (Nishimura, 2012, p. 12).

The transformer issue was more or less solved. But that was only one part of the total system. The electric light in the form of the incandescent lamp would be the second issue that needed to be addressed. Westinghouse started acquiring essential patents—like the Sawyer-Man patents for incandescent lamps that he obtained by acquiring Consolidated Electric Co., a company that was controlled by Houston-Thomson Electric Co.[97]

---

[97] For more details see: B.J.G. van der Kooij, *The Invention of Electric Light*. (2015)

Once Westinghouse secured the Sawyer-Man patents, the actual job of making an Edisonized Sawyer lamp fell to Oliver Shallenberger, one of Westinghouse's bright young men…Once Westinghouse determined to stake everything on alternative current, he sent Shallenberger off to study at the feet of the master, Alexander Lodyguine, in St. Petersburg. By the time he returned a year later, Westinghouse had acquired the United States Electric Light Company and thus controlled the patents and licenses to all the attenuated treating methods of carbon filaments, including Maxim's patent on an improved Sawyer-Man (McPartland, 2006, p. 306).

A third component in the total AC system was the electric motor. And that is where Nicola Tesla, who held a presentation on "A new System of Alternating Current Motors and Transformers" in May 1888 for the AIEE, came on stage. By July 1888 Westinghouse had licensed Tesla's motor, with a generous royalty based on the total annual horsepower developed by all Tesla motors. The AC motor developed by Tesla was the last component in the total system Westinghouse had envisioned. By 1889 Westinghouse had hundreds of AC installations up and running or under construction.

Westinghouse Electric Co.

In March 1886 the electrical department of the Union Switch and Signal Company became the Westinghouse Electric Company, capitalized at $1 million. And the company was soon seen as a dominant player in the AC-electricity business. The fact that the company was able to handle big electrification projects (like the Columbian Exhibition and the Niagara project) contributed to this development.

Figure 144: Winding railway motor armatures at Westinghouse's factory (1887)
Source: http://memory.loc.gov/ ammem/papr/west/westwind.jpg

Columbian Exposition (1893): Westinghouse Electric Company got the order for the lighting equipment at the Columbian Exposition at Chicago in 1893 (Figure 145). This event was to celebrate the four hundredth anniversary of the discovery of America. The contract called for 92,000 lights to illuminate the fair.

On May 23, 1892, the Westinghouse Company took the lighting contract at a price much below the bid made on behalf of the Edison General Electric Company, its only serious competitor. The story is that the savings to the Exposition Company was something like $1,000,000, which may well have been, as the unit prices were about as one to three. The Edison General Electric Company counted on its strong patent situation, and Westinghouse set high value on the advertising element. His company lost money directly, but its technical success had a great effect on the Niagara Falls contract then pending, and on the whole struggle between direct current and alternating current, and it is hard to exaggerate the world importance of that struggle (Prout, 1921, p. 134).

Figure 145: World's Columbian Exposition, Chicago, 1893

Source: http://sites.roosevelt.edu/hbarnett/files/2013/06/Palmer-book-cover.jpg

Now Westinghouse had the contract, but he was still facing Edison in the patent struggles over the incandescent lamp. So he had to use a noninfringing lamp design that soon became manufactured in volume (Figure 146). As the courts had decided that the Swayer-Man design did not infringe on the Edison patent, Westinghouse decided to use the "stopper-lamp" design for the lamps at the fair.

Figure 146: Westinghouse employees manufacturing stopper lamps for the Columbian Exposition (1893)

Source: IEEE Global History Network, www.ieeeghn.org

Quite apart from the lighting plant, the Westinghouse Company showed at the World's Fair a complete polyphase system. A large two-phase induction motor, driven by current from the main generators, acted as the prime mover in driving the exhibit. The exhibit, then, contained a polyphase generator with transformers for raising the voltage for transmission; a short transmission line; transformers for lowering the voltage; the operation of induction motors; a synchronous motor; and a rotary converter which supplied direct current, which in turn operated a railway motor. In connection with the exhibit were meters and other auxiliary devices of various kinds (Prout, 1921, p. 139).

Niagara project (1895): Another big project was the electrification of the Niagara waterfalls. This would become one of the most famous early hydroelectric power stations. The Niagara Falls, with their enormous flow rate of water (ca. 110.000m³/minute), were already used for the creation of hydroelectric power (1881, DC electricity) on a limited scale. But in 1890 the Cataract Construction Company, created for the purpose, was asking for new and larger designs for the utilization of power from the falls. In 1890 the International Niagara Commission advised the creation of a single, large power plant and decided that the AC electricity should be used for its transmission and distribution. The commission issued invitations for proposals in three categories: power development; transmission and distribution; and a combination of the first two categories. Seventeen projects were submitted, of which three were dismissed as not complying with the terms of the invitation or insufficiently complete to warrant judging. For transmission and distribution, seven of the fourteen proposals were for electricity. Five were DC and two were AC. One of the AC proposals was for polyphase, that is, two separate currents or waves of electricity (since AC takes the form of a sine wave) out of phase. Neither Edison nor Westinghouse submitted a project to the commission. In December 1891 the Cataract Company issued an invitation to six companies for design and construction of the electrical installation for the Niagara project. Both General Electric and Westinghouse submitted their bids for the Niagara project in two stages. General Electric's first-stage bid was submitted in the fall of 1892. It was for DC locally and AC to Buffalo. The first-stage Westinghouse bid was submitted in December. It was for polyphase AC. In March 1893 both companies submitted final bids. Both now proposed polyphase AC. The two proposed systems were virtually identical, except that the Westinghouse system was two-phase and the GE system three-phase. But they were not accepted.

Figure 147: Generator installation at Adams Station power plant at the Niagara Falls project
Source: IEEE Global History Network.
http://www.ieeeghn.org

> *While the Westinghouse Company proposed two-phase generators, the General Electric Company recommended a straight three-phase system. After examination by the engineers of the Cataract Construction Company, the tenders of both manufacturing companies were declined, and the Cataract Construction Company instructed its engineers to prepare an alternative generator design…In again asking for bids, the Cataract Company said, referring to the alternative generator design prepared by its engineers, "any alterations that you may propose in this design will be carefully considered and if acceptable will be appreciated in placing the contract"* (Prout, 1921, pp. 148, 150).

The new request for proposals elicited an irate letter from Westinghouse, who may have suspected that the Niagara Company was contemplating designing or even building the apparatus in-house, and was merely wanting the benefit of his company's ideas, free of charge. Westinghouse Company won the contract for the three 5,000-horsepower generators. General Electric got the contract for the transformers and transmission. On August 26, 1895, the first power was produced from the Niagara Falls generators. Originally the electricity was meant to power the Buffalo streetcar system, but that turned out quite differently. The availability of abundant cheap power spawned an entirely new industry in Niagara Falls—the electrochemical industry—that had power requirements right from the start that accounted for virtually the entire supply. The first of these companies was the Pittsburgh Reduction Company (later renamed Aluminum Company of America). This company was founded in Pittsburgh in 1886 but transferred operations to Niagara Falls during this period because of the prospect of cheap and reliable electrical power. The second electrochemical industry, the Carborundum Company, also transferred from the Pittsburgh area.[98]

These two projects gave Westinghouse the edge he needed to become a dominant player in the AC-electric industry. The technology and products that allowed the Westinghouse Electric Company to enter into the electrical-equipment business were, for the most part, not invented or developed by George Westinghouse himself. Instead, he organized competent engineers and their patents to become a big business in this area—for example William Stanley with his improvements to the transformer; Nikola Tesla, with his induction motor; the Swiss engineer Alfred Schmidt, who worked on dynamos; Oliver B. Shallenberger, with his AC meter; Benjamin G. Lamme, who joined the company in 1888 and later

---

[98] Text on the Niagara project based on: William S. Dietrich III: George Westinghouse, the mystery (2006). http://www.pittsburghquarterly.com/index.php/Historic-Profiles/the-mystery-of-george-westinghouse.html. (Accessed December 2014)

become a chief engineer; and Charles F. Scott, who soon became a professor at Yale University.

George Westinghouse was intimately involved at every stage of the game. His technological versatility, his entrepreneurialism, his manufacturing capability, and above all, his force of personality carried the day. But he lacked expertise in two major fields: marketing and finance.

> *In shifting his attention from his railroad inventions (the air-brake and signaling systems) in the mid-1880s, George Westinghouse wisely hired a number of electrical inventors and engineers, including William Stanley and Nikola Tesla, and hence ensured that he had ready access to new technological developments. To manufacture the electrical systems developed by his inventors and engineers, Westinghouse built factories in Pittsburgh and Newark. However, his weakness was in marketing; Westinghouse depended on a small sales force working on commission out of offices in six or seven major cities, and he insisted on closing many of the major deals himself. Westinghouse also bankrolled the early development of his electrical company himself, which meant that he never developed the strong relationship with a banking house needed to float bond and stock issues or to help finance central stations purchasing his equipment. Consequently, again like the Edison organization, the financial crisis of 1890 found Westinghouse bankrupt, and he was only able to reorganize with the help of financiers August Belmont and Henry L. Higginson* (Carlson, 2013, p. 77).

After the great success achieved in 1893, the company developed its electrical systems—particularly turbo generators—by acquiring licenses for the US patents of Person's turbine in 1895. The company then began the electric-train business. The Westinghouse Electric & Manufacturing Company became the second largest electric company (the General Electric Company (GE) was the largest) and aggressively entered into foreign markets prior to World War I. At the turn of the century, the various Westinghouse companies were worth about $120 million and employed approximately 50,000 workers. By 1904 there were nine Westinghouse manufacturing companies in the United States, one in Canada, and five in Europe. By 1907 sales had reached $35 million.[99] The financial panic of 1907 struck Westinghouse hard. Westinghouse lost control, and the East Coast bankers put their own man in to rein in the expansionist bent of the founder. By 1910 Westinghouse was no longer on the board.

---

[99] This would be more than $600 million in 2010, using historic opportunity costs. Source: http://www.measuringworth.com/uscompare/relativevalue.php

In 1907 came the tragedy of Westinghouse's life. The great panic caused the failure and receivership of the Electric Company, the Machine Company, and some minor companies, but did not affect the Air Brake Company or the Union Switch & Signal Company. A reorganization was eventually brought about, based upon a brilliant project devised by Westinghouse, but the actual control of the Electric Company passed out of his hands, and in less than four years he ceased to have any official relations with the company (Prout, 1921, p. 19).

Four years after his retirement, on March 12, 1914, George Westinghouse died in a wheelchair. Not surprisingly, close at hand were design drawings for an electric wheelchair he was working on. At the time of his death, George Westinghouse controlled more than 15,000 patents, 314 of which he had invented himself. A 1921 biography puts the total number of Westinghouse-affiliated companies at 104. When he died, he had more than 50,000 workers in his direct employ. He had founded and built the Pittsburgh suburb of Wilmerding. Despite his death, his companies continued to change the world (Imerito, 2013).

Battle of currents: DC versus AC

The development of electrical systems was characterized by fierce competition. One battle was fought between those who believed in the "isolated plant" system (and had a vested interest in it) versus those who believed in the "central plant" system (and had a vested interest in that system). But there was another battle going on between those who believed the future for electricity was in direct current (DC) and those who believed in alternating current (AC). With the electric dynamo an alternative source of electric energy, a different form of electricity had become available. The use of voltaic batteries had limited the electric energy to DC systems. But with the dynamo, AC became available.

The arguments pro and con were exchanged in the highest echelons of science, such as William Thomson/Lord Kelvin who, in 1881 in his address "On the Sources of Energy in Nature Available to Man for the Production of Mechanical Effect" stated:

> *High potential—as Siemens, I believe, first pointed out—is the essential for good dynamical economy in the electric transmission of power. But what are we to do with 80,000 volts when we have them at the civilized end of the wire? Imagine a domestic servant going to dust an electric lamp with 80,000 volts on one of its metals? Nothing above 200 volts ought on any account ever to be admitted into a*

house or ship or other place where safeguards against accident cannot be made absolutely and forever trustworthy against all possibility of accident (Thomson, 1881).

Thomas Edison promoted the DC systems with 110 volts, but they had a limited range of application (within a one-mile radius) due to the loss of power over longer transmission lines.[100]

George Westinghouse advocated the AC system (Figure 148) because it could be operated with higher voltages over longer distances, using step-up transformers to increase the voltage. At the end of the distribution line, the voltage would be transformed to the lower voltage by the step-down transformer. This had the practical significance that fewer, larger generating plants could serve the load in a given area. Large loads, such as industrial motors or converters for electric-railway power, could be served by the same distribution network that fed lighting, by using a transformer with a suitable secondary voltage. When Tesla introduced a system for alternating current generators, transformers, motors, wires, and lights in November and December 1887, it became clear that AC was the future of electric power distribution, although DC distribution was used in downtown metropolitan areas for decades thereafter.

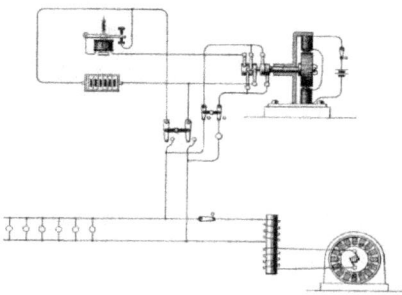

Figure 148: Westinghouse's early AC system (1887)
Source: Wikimedia Commons, USPTO: US-patent 373.035

Edison versus Westinghouse

George Westinghouse, after becoming rich and famous with his invention of the railroad air brake, went into the electric business in 1885 and became a major player with the purchase of the United States Electric Lighting Company in 1886. He preferred an AC-based lighting system. He became the direct competitor of Edison Electric Lighting Company, which used a DC system.

---

[100] In contrast to AC electricity, which uses the outside of a wire to carry the current (the "skin effect"), the electrons in DC electricity travel within the copper wire. As the wire has a resistance, this leads to a drop in the voltage over the distance of the wire. And, as a thin wire has a higher resistance than a thick wire, much more (expensive) copper wire is needed for DC systems than for AC systems.

Westinghouse's AC system

Westinghouse was instrumental in two developments. First, he advanced the art of transportation by the invention and development of the air brake. Second, he advanced the manufacture of electric power by the development of the use of the alternating current in the distribution and application of electricity (Skrabec, 2007, p. 102).

> He [Westinghouse] was a stocky man: blunt, dynamic, a bit of a fop, with a walrus mustache and a taste for adventure. Westinghouse was a hard-driving businessman, but the antithesis of a robber baron. He did not believe it was right or even necessary to bribe politicians or cheat the public to be successful...Unlike Edison, Westinghouse did not go into the electrical business by manufacturing and marketing his own inventions. Instead, he bought up available patents and hired a skilled staff of engineers to work on making practical improvements

(Moran, 2007, pp. 47-48).

It was William Stanley, the inventor of a specific self-regulating dynamo, who designed an AC system using transformers. This system was demonstrated to the public on March 20, 1886, in Great Barrington.

> Westinghouse supplied Stanley with a complete laboratory, paid him a generous salary, and covered all expenses up to $200 a month. He also gave Stanley one-tenth of the stock in the soon-to-be-formed Westinghouse Electric Company. In return Stanley agreed to surrender his rights to all commercial products he might invent...On March 20, 1886, Stanley held a public demonstration. Copper wires were strung from the old rubber mill to the center of town, a little less than a mile...In the early evening, Stanley illuminated thirteen stores...two hotels, two doctor's offices, one barbershop, and the telephone and post offices. The era of alternating current had begun. Electrical history was transformed...Unlike direct current, customers were no longer required to live within a few hundred yards of the central station to obtain electrical service. The Westinghouse system was an immediate success. Within a few months, twenty-five orders for new central stations had been taken. By the end of two years, Westinghouse had installed his system in 130 cities and towns. Edison sales agents became alarmed...

(Moran, 2007, pp. 49-50).

It was the United States Electrical Lighting Company, who owned the right to the Sawyer patent on an incandescent lamp, that hindered Edison from obtaining his US patent № 223.898, and he decide to sue that company for infringement. Westinghouse wanted to enter the incandescent-lamp market and bought the rights to the Sawyer patent. That made him a participant in the patent fight. So, in a way, it became Thomas Edison against George Westinghouse.

Edison decided to fight Westinghouse

Edison began a lobby campaign in several states legislatures to pass laws limiting the voltage permitted for power lines. He testified in hearings for committees in the Ohio and Virginia State Senate. But he did even more than that. By 1887 he had embarked on a campaign to discredit AC as too dangerous to be considered for lighting systems. He asked his managers to collect information about fatal accidents involving AC. In February 1887 he published a red booklet called *A Warning from the Edison Electric Light Co.* in which he gave a response to the "misstatements and misrepresentations put forth by these companies..." (the companies being Consolidated Electric Light Co., Westinghouse, and Thomson-Houston Co.). He warned readers about the dangers of alternating current and pointed to the glorious record of his own system. Among other accusations he attacked Westinghouse's entrepreneurial activities by highlighting his involvement in financial activities. In an appendix titled "The Westinghouse Stock Boom," he concluded: "Stockholders and others interested can form an opinion of whether $3,000,000 is a just estimate of the commercial value of the two patents" (Edison Electric Light, 1887).

The booklet's articles expounded upon the dangers of electricity with headings such as "Horrible Death of a Lineman," "The Wire's Fatal Grasp," "One Martyr More," "Wire Has Another Victim," "The Electric Murderer," "Another Lineman Roasted to Death," "Electric Wire Slaughter," "Again a Corpse in the Wires," "Death's Riot," and "Electric Wires Add to Their List of Victims." All of the articles covered the same accident (Leupp, 1918).

> *This line of agitation at first appeared to come almost wholly from inexpert or at least nonprofessional sources; but presently arose one Harold P. Brown, an electrician by calling, who, not content with denouncing the survival of overhead wires in a great city, made the alternating current itself, wherever found or however used as a public utility, an object of attack* (Leupp, 1918, p. 145).

Brown—who had worked for Western Electric Company and Brush Electric Company, maintaining arc streetlights in Chicago from 1879 to 1844—had interesting connections. After his work in Chicago, he went into business, became a consultant, and obtained several patents (US Patents №. 330.465, №. 352.035). Brown was given full use of Edison's research labs in New Jersey. Edison authorized a series of experiments in his laboratory—assigning his employee Arthur Kennely to give assistance—in which animals would be killed to demonstrate the dangers of AC. Then in the presence of newspaper reporters and invited observers, researchers forced dozens of stray dogs and cats onto the grid, where they were instantly killed (Brandon, 2009, pp. 70, 74). These violent experiments that Brown performed created massive publicity (Figure 149).

Figure 149: The execution of a horse in an Edison Laboratory
Source: http://seaus.free.fr/spip.php?article500

The atmosphere became, for a while, thick with the personalities, including charges of interested motives and even of bribery and fraud, volleyed back and forth between the champions of the respective systems…Interviewers pursued Westinghouse wherever he went, trying to lure him into some explosive utterance against Thomas A. Edison, the chief exponent of the continuous current, which might produce a personal collision between the two inventors, and thus set free a fund of spicy "copy."…Even the most sober of the great periodicals were drawn into the controversy. An article on "The Dangers of Electric Lighting," arraigning the alternating current, by Thomas A. Edison, appeared in the North American Review, and "A Reply to Mr. Edison," by George Westinghouse, in the next month's number (Leupp, 1918, pp. 147, 148, 150).

But it was not only the press that acted. The mayor of New York City was also under pressure to do something.

> *In the midst of the turmoil, Hugh J. Grant succeeded to the mayoralty, and his office became the storm-center of a tremendous struggle which lasted about two years, and was punctuated at intervals by court orders, injunctions, and counter injunctions, and by raids made upon the overhead wires by gangs of municipal employees under orders to cut away all that were improperly insulated, obstructively hung, or otherwise liable to be dangerous…As a result, the great city [New York] was left almost in darkness at times, as arrangements for going back to lighting the streets with gas were not easily perfected*
> (Leupp, 1918, pp. 150, 151).

Westinghouse reacted as well.

> *Westinghouse did not appreciate the free publicity. He considered suing, but forbore doing so. But in magazine articles, he and Edison—two of the supreme egos of nineteenth-century American business enterprise—went head to head.*

Edison reiterated the danger of AC current and argued for outlawing high voltages, that is, above several hundred volts, which would have wiped out AC's mechanical and commercial advantage (of high-voltage transmission). Westinghouse's argument was that the only voltages that mattered were in buildings, where people might possibly come into contact with them, and where AC voltages were reduced to 50 volts. (He further pointed out, in a delightful nonsequitur, that because the transformer worked on an induction principle, there was no direct contact between the high tension current in the transmission lines and the building current.) For his part, Westinghouse proposed outlawing building currents above 100 volts (DC building current was 110 volts)[101] (Foran, p. 14).

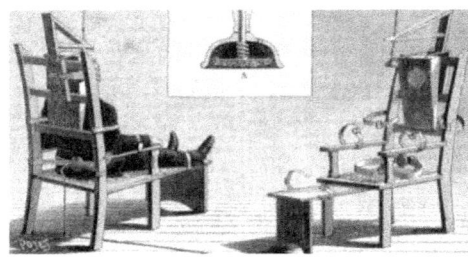

Figure 150: The execution of William Kemmler (1890)

Source: http://seaus.free.fr/ spip.php?article500

During the gruesome experiments with animals, the discussion turned to using the electric chair to execute criminals condemned to the death sentence (Brandon, 2009, pp. 78-80). The dentist Alfred Southwick first expressed the idea. (He had witnessed an intoxicated man touch a live electric generator and die instantly from the accident.) The state of New York considered his idea and started a legal process introducing this method as an alternative to the current mode of capital punishment: hanging. Edison, being a renowned scientist and expert on electricity, was asked for advice. Being opposed to the death penalty, Edison declined, but at a second request, he responded differently.

The best appliance in this connection is, to my mind, the one that will perform its work in the shortest space of time, and inflict the least amount of suffering upon its victim. This, I believe, can be accomplished by the use of electricity, and the most suitable apparatus for the purpose is that class of dynamo-electric machines which employs intermittent currents. The most effective of these are known as "alternating machines" manufactured principally in this country by George

---

[101] Westinghouse countered by asking lawmakers to make laws limiting the voltage in homes to less than 100 volts. This early battle became the basis for the standard 110 volts in the United States, while most of the rest of the world uses 220 volts in households and offices. http://www.electric-history.com/~zero/005-Electricity.htm

Westinghouse (Moran, 2007, pp. 74-75).

Later, Harold P. Brown together with Arthur Kennelly designed and built an electric chair operating on AC, based on Southwick's concept (Figure 150) (Brandon, 2009, p. 73). William Kemmler was the first person to die by electric execution, and on August 6, 1890, the next day's papers headlined: "Kemmler Westinghoused" (Moran, 2007, p. XX).[102]

> *Elihu Thomson stood by the side as Westinghouse and Edison fought the battle. When asked by his management to defend the safety of AC, he wrote, "I have no panacea—for all the ills which may follow the use of high potential currents under conditions usually found in large cities. I can no more say how to make electricity safe in such cases than I can say how to make railroad travel safe, or how to make steamship travel safe, or how to make the use of illuminating gas safe, nor the use of steam boilers safe. no improvement of our modern civilization has ever been introduced but that involved considerable risk*
> (Carlson, 1995; Klein, 2010).

The competition between the Edison DC system and the Westinghouse AC system became fierce. It was not just about patent infringement, anticompetitive legislation, and public-relations schemes to discredit the opposition and frighten its customers. It was also the battle between arc-light systems and incandescent-light systems. Already, within two years of its advent, the Westinghouse Electric Company's rival system had sold more central station plants than all other DC companies combined. Westinghouse Electric's growth was extraordinary—from sales of $800,000 in 1887 to $3 million in 1888. In 1891 AC systems had a greater than 50 percent share in the field of lighting. The alternative AC system proved to be more economical (investment wise and operational wise) then the DC systems. The newly developed AC-DC converter—the Bradley rotary converter, a device that combined an AC induction motor with a DC dynamo—could even combine (local) DC networks with AC networks. And AC systems changed from single-phase systems to triple-phase systems (Paul A. David & Bunn, 1988, pp. 173-179).

The end of the battle

The battle of the currents was more than the described conflict between Edison and Westinghouse. It was the conflict between the vested interests in different systems of distributing electricity. And the winner was AC, simply because the nature of the technical properties of AC electricity were better suited to the transportation of electricity over long distances. An

---

[102] For details see: (Reynolds & Bernstein, 1989).

important event in that turning point was the Electric Exhibition that took place in Frankfurt am Main in 1891.

But the technical change to three-phase AC took some decades more, as there was much invested in the installed base of the DC system. Soon devices that could convert AC to DC were invented—such as the rotary converter, developed by Charles Bradley and Friedrich Haselwander. Thus, existing single-phase equipment (AC and DC) could be coupled with the new system. The concept of a universal system completed the transition from the era of light to the era of light and power (Hughes, 1993, p. 122).

> *The era of rivalry between technologically distinct systems was brought to a close within six years of its commencement. Yet, the question of the superiority of one form of current over the other remained unresolved within the engineering community…While perhaps true in 1890, the superiority of DC was more doubtful a few years later. In 1891, Oscar Muller and the Swiss firm of Boveri and Co. demonstrated that polyphase current could be transmitted the 110 miles from Lauffen on the upper Neckar River to Frankfurt-am-Main, Germany…Equally impressive was the Westinghouse polyphase system exposition (including the rotary converter) at the Chicago World's Fair in 1893. These events made it evident that lighting was no longer the only factor to consider when discussing the load and efficiency characteristics of AC and DC systems With the extension of AC to power users, and to traction users as a result of the invention of the rotary converter, the decision between AC and DC came down to which of them could distribute power over a distance most efficiently and cheaply*

(Paul A. David & Bunn, 1988, p. 186).

Another question that could be asked is why Edison chose the tactics that he did. In trying to find an answer, there is both the person and the circumstances to consider.

> *Edison's single-minded determination to hammer home the deadly nature of AC began to echo in all his public statements…By all accounts Edison was a stubborn man. He had a strong belief in himself, always thinking he could succeed where others had failed. Often iconoclastic and usually audacious, Edison tended to dismiss the opinions of others, especially when they clashed with his own…His hatred for George Westinghouse, his ego-investment in the industry he created, and old-fashioned pride all made it difficult, perhaps impossible for Edison to act as the shrewd entrepreneur he was. His cutthroat competition with Westinghouse would test the limits of Edison's character, leading him to risk his reputation as a national hero and ultimately to betray the public's trust*

(Morton Jr, 2002, pp. 58, 61-62).

Also, one has to realize that by 1885 Edison and his coworkers were not in sole charge of the Edison companies, due to constant reorganizations. The financiers behind the company manned the new boards. The consolidation of the Edison companies into General Electric in 1889, with the backing of Deutsche Bank, the Algemeine Elektrizitäts Geselschaft (AEG), and Siemens & Halske, had left Edison with $1.75 million[103] in cash and 10 percent of the shares in General Electric.

> *The inventor saw in this a welcome opportunity both to extricate himself from the worries and distractions of managerial and financial responsibility for the manufacturing business, and to raise sufficient capital to place his laboratory on firmer financial foundations…On February 8th [1890] he wrote Villard not to oppose his "retirement from the lighting business, which will enable me to enter into fresh and congenial fields of work."* (Paul A David, 1991, pp. 95-96).

By 1890 Edison retired from an active role in the company. The propaganda war against the Westinghouse AC system, which had been brought to its peak in the midst of the consolidation negotiations, rapidly wound down in 1889 (Paul A David, 1991, pp. 94-96). Edison focused on work in his new laboratory in West Orange, New Yersey, that he started in 1887. Here he continued patenting his next inventions (more than 50 percent of his 1,093 patents came from this laboratory)—inventions like motion-picture cameras, improved storage batteries, and an improved phonograph.

---

[103] This amount would be equivalent to more than $43 million in 2010, calculated on the basis historic standard of living. Source: Measuring Worth at http://www.measuringworth.com/uscompare/relativevalue.php

Conclusion

Anno 2015 nobody, flicking on an electric light, starting the dishwasher, using the elevator, preparing an espresso coffee, nor any other moment even pays a second thinking about what made this all possible. Electricity is taken for granted, electric technology is the unnoticed part of our daily lives. But it took nearly two centuries to make all that possible. And the foundations were laid in the nineteenth century.

In the preceding segment we have looked at the General Purpose Technology of 'electricity'. We have identified its three major clusters of innovations; 1) the cluster around the *basic-innovation of the DC-electromotor*, 2) the cluster around the *basic-innovation of the electric dynamo*, and 3) the cluster around the *basic innovation of the induction electromotor*. Our major theme was the quest into the Nature of Innovation. More specifically we focused on innovation related to electricity. How could these three basic innovations revolutionize the world we live in today? Creating an utter dependence of societies on a single phenomenon called 'electricity'. Giving us all the comforts of electric light and the electric domestic appliances. Facilitating our modern tele-communications and fulfilling our information needs. What did it make happen to be?

Just reflecting on the massive social changes that originated from the contributions of so many, willing to devote their creative efforts in changing the world, we will try and wrap up this case study with the following interpretations of our observations.

Human curiosity, ingenuity, and competition

The discovery and application of the electromotive engine (more popularly called the "electric motor") certainly could be an outstanding invention if it had been the act of one person. But that was not the case, as shown before. The results came about because of a range of discoveries made by many experimental scientists and engineers that started with electromagnets and ended with the creation of the range of AC and DC motors we use today. It took a while to progress from Volta's pile (ca. 1800), Davenport's electromotive engines (let's say 1830), to Doblovosky's and Tesla's induction motor (ca. 1890). The progression even called for a detour to replace the battery (the "wet cell") with the electric dynamo (the "dry cell").

But when electricity could be easily generated in abundance in the second half of the nineteenth century, a range of events occurred—events that resulted in the development of the electromotive engines themselves as well as the range of stationary and mobile applications for the electromotive engine. From those simple distribution systems for DC electricity, the more complicated poly-phase AC distribution networks were developed. This, in turn, made the use of electric motors in industrial applications more feasible—from the motors for powering ventilators, lathes, and sewing machines to the electric-powered streetcars and elevators in high-rise buildings. Electric power, in combination with steam and waterpower as "prime movers," complemented and replaced the steam engines in industrial power networks. From the central "shaft and belt" power distribution system grew the individual, electrically powered tools and machines. Electricity, especially the alternating current electricity that could be easily transported over large distances, became a general power source for light and motor applications.

Curiosity in the nature of lightning

It all started with curiosity and with people asking themselves questions. Why did things happen as they did? Why did the frightening lightning in the sky occur? Why did it have such a loud noise and such deadly force? Could we catch hold of it and bring it to earth? Could we imitate it when we rub a cat's fur against an amber stick? Combined with a creativity to experiment, the inquisitive looked for solutions. What happens if we manipulate a frog's legs? What is the result of stacking different metals in an acid solution? In short, many curious people asked themselves questions about the nature of lightning and how it could be used.

Slowly, insight about the nature of lightning was gained. An insight created by the collective curiosity of all of the inventive people spending their time and money experimenting. Without understanding the reasons why, the "gentlemen of science" and "common engineers" in Europe started exploring the phenomena related to the newfound electricity. They discovered, step by step, the properties of electricity, including *electromagnetism*. It was electromagnetism that provided the missing link between "electricity" and "movement." Not only could a current from a bank of voltaic cells spark into light, as Humphry Davy demonstrated, it could also move a compass needle, due to magnetic properties, as Hans Christian Oersted noted. Their work was discussed broadly in the scientific communities, like the Royal Society of London and the Academy of Sciences in Paris. Now other scientists, in turn, soon demonstrated the principles to other people from a different, often less privileged, background, which sparked their curiosity and created a stimulus to explore new findings. An example is Faraday, who discovered that magnets could induce electricity, and also people such as Sturgeon and Henry who started experimenting with the new phenomenon of electromagnetism, creating artifacts (later called electromagnets) that realized linear and rotational movement. They showed that this basic mechanism was a strong force that could hold considerable weight.

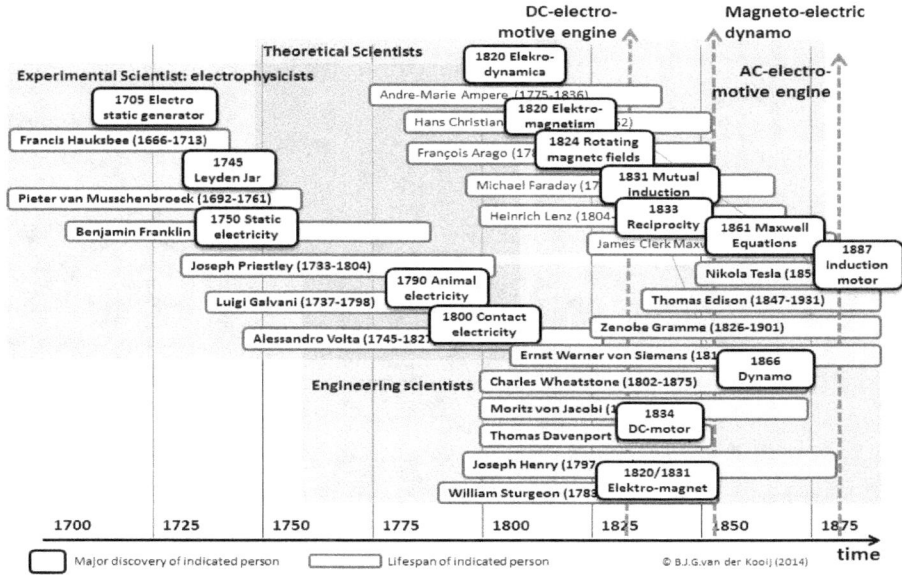

Figure 151 : Overview of scientific discoveries regarding electricity

Source: Figure created by author

Ingenuity

Volta's "wet cell'" was an important discovery that showed that "movement" could be realized by using electricity created chemically. Copying the steam-cylinder concept into an electromagnet application proved to be a dead-end technology for the linear motor. It was going to be the rotative movement created by electromagnetism that would have a future.

Rotative movement by electromagnetic force was an impressive phenomenon that excited many and encouraged further experimentation. Experimental scientists created electric motors that powered a small boat on the river Newa in St. Petersburg and that powered electric trains in London. The DC electric motor was born; it more or less worked, and it was expected to have broad applications. But further application of this electric motor was obstructed, as it was hindered by the limitations of the same wet cell that made the dynamo-electric machine (as the electric motor was called) possible.

The solution was as simple as the discovery of the dynamo effect. People such as Faraday and Lenz had discovered that, as electricity could create movement (the dynamo-electric machine or electric motor), the

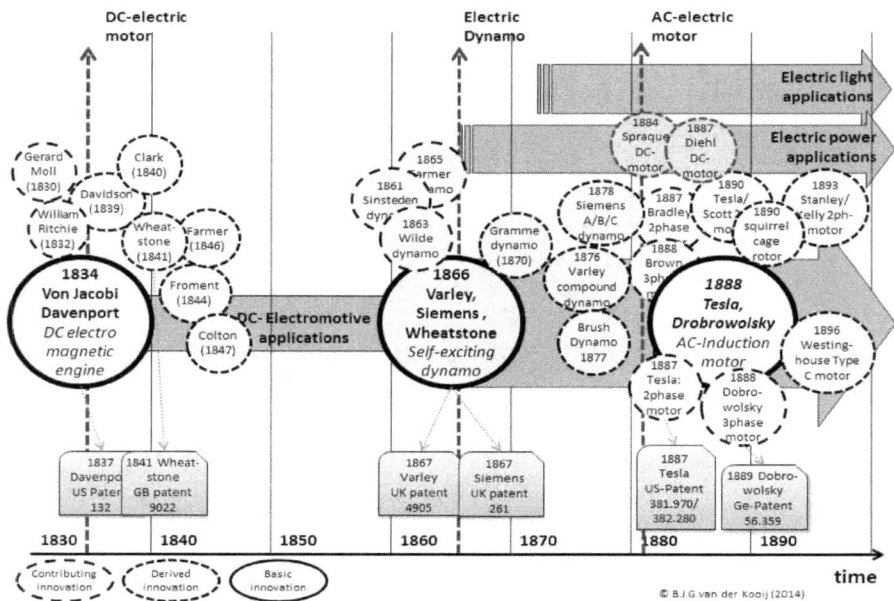

Figure 152: Overview of clusters of innovations around the electromotive engines

Source: Figure created by author

process also worked the other way. Rotative movement of a coil in a magnetic field could create electricity. It would be the antagonist of the dynamo-electric machine—the electrodynamic machine was followed by the dynamo-electric machine (or dynamo as it later became named)—that lifted the obstruction. This "dry cell" that could generate electricity in abundance was the newfound source of electricity.

Then, in the *era of light*, electric lighting created the "market pull" that complemented the "technology push." People were fascinated by the demonstrations of the new arc lights that were lighting streets and theaters. And when the incandescent lamp was developed, enormous lighting markets fueled further developments. From simple, locally based DC-electricity distribution networks to larger municipal DC-electricity networks, electricity became available on a larger scale. And when the discovery of the induction effect made the AC-induction motor feasible, the large three-phase distribution networks brought electricity nearly everywhere. It would fuel the *era of power*.

Competition

Many individual scientists, inventors, and engineers contributed to the total development. Some contributions had a small impact; others had an influence that changed the course of electrical development. Although mostly dominated by the technological potential of electricity to transport power, the developments took place in the specific context of the nineteenth century—a context that was dominated in the United States by its capitalism, resulting in massive business creation and monopolies. Electrical progress happened in a totally different context in many European countries, each with its own character; yet similar developments took place, also leading to massive business generation and giant companies.

The context may have been different between the Old World of today's Europe and the New Word of North America, they had one element in common; competition for survival. Certainly the capitalist system has been creating a highly competitive structure for individuals and organizations to earn an existence and survive. But also the more socially oriented European system had competitive elements where individual and organizations were faced with. Both systems, each in its own way, was about the Darwinian "survival of the fittest". The fittest technology, the fittest company, the fittest products. A process in which technologies, companies, infrastructures, systems and products were created, pioneered, matured and died. A process of business cycles with its creative synthesis and creative destruction.

Societal change induced by technical change

What can be observed from our exploration into the General Purpose Technology of 'electricity'? How can we interpret the relations between social change and technical change? Let's try and identify some of the characteristics.

Second Industrial Revolution: "Power to the people"

The European Revolutions of 1848 mark the different periods of social, technical and political change. The first half on the nineteenth Century was still dominated by the *First Industrial Revolution*. The second half of the nineteenth Century was to be dominated by the *Second Industrial Revolutution*. First in England where the *Great Victorean Boom* (1850-1873) took place. Then, after the dust had settled on the madness of times that ruled Europe for so long and the new political structures were in place, a period of relative peace commenced, in which economies and societies bloomed. It was the time for the *Belle Epoque* in Europe (1871-1914) . The same goes for the US where, after the end of the Civil War, the *Gilded Age* (1865-1905) illustrated the prospering country. (Figure 153)

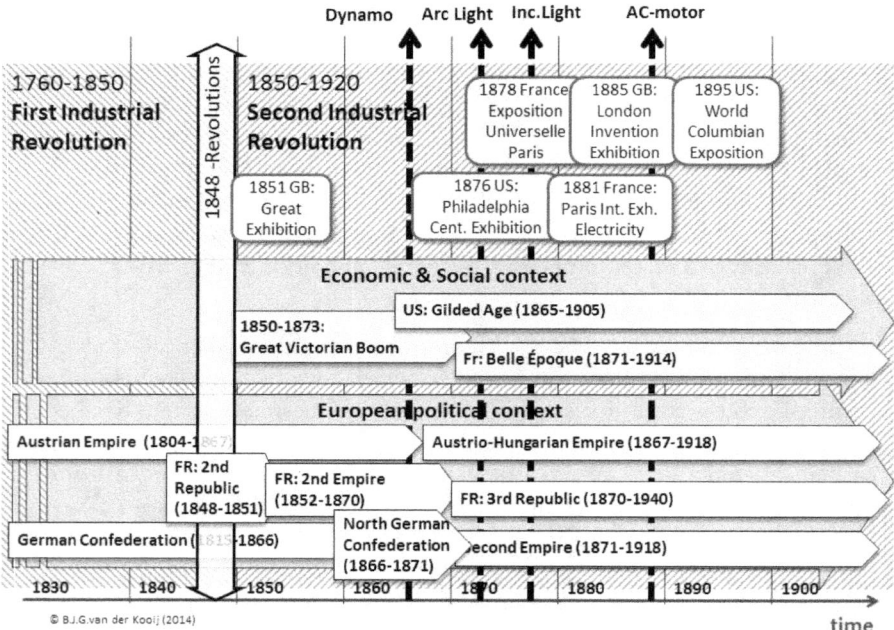

Figure 153: The context for the Second Industrial Revolution

Source: Figure created by author

Public awareness

Imagine a person from the preelectric era being confronted with a new thing called electricity. The wonders of electricity were shown to the astonished public at exhibitions such as the Great Exhibition of the Works of Industry of All Nations (the Crystal Palace Exhibition) in London in 1851, drawing more than 6 million visitors. The 1876 Philadelphia Centennial Exposition drew nearly 10 million visitors, who watched the wonders of steam technology, Alexander Bell's telephone, and the Farmer-Wallace electric dynamo. The *Exposition Universelle* in Paris in 1878 had 13 million visitors, and the International Exhibitions of Electricity in Paris in 1881 and the London Inventions Exhibition in 1885 had 3.3 million visitors.

They all created a massive public interest in the electric light—from arc light to the incandescent lamp (in addition to other miracles such as the telegraph, telephone, and phonograph). The early demonstrations of the arc light in Philadelphia, Holborn Street (London, England), and *l'Opera* (Paris, France) and demonstrations of the incandescent lamp at Menlo Park and Pearl Street (New York, United States) excited the public and spurred the early entrepreneurs into action. The International Electro-technical Exhibition in Frankfurt in 1891, the World Columbian Exposition in Chicago in 1893, and the Niagara project in 1895 where milestones in the development of AC systems. From then on, electricity's penetration into society became a fact.

Pervasiveness

The preelectric person would certainly have problems envisioning the role of electricity as it developed into in the twenty-first century. It was hard to foresee those developments—like use in the private environment of the home, where dozens of AC motors would be supplied from the access point to the National Grid[104]. Appliances such as washing machines, dishwashers, refrigerators, coffee machines, blenders, electric clocks, and private-community gate access, just to name a few, are now powered by AC-electric motors.

How could one foresee that individual mobility (trains, tramways, and cars) would depend on electricity as well as? How could one foresee a working environment dominated by tools and machines using AC motors? Examples are handheld tools and fixed tools (lathes, milling, and drilling machines) and specific stationary machines that are used for an abundance of applications (i.e., packaging foods, filling beer bottles). And, finally, how

---

[104] This the name for the high voltage electrical power transmission system

could our preelectric person foresee that living conditions would be influenced by the use of electricity at home: electric cooling, heating, and cooking?

Power supply infrastructure

The electricity that powered all of these applications would be supplied by a system of interconnected electricity-distribution networks—regionally, nationally, and internationally. Electricity-generating plants are powered by hydroelectric, nuclear, gas, and coal as "prime movers"—thus creating a society that totally depends on one, single phenomenon: electricity. This is a precarious existence when one considers the problems caused by the Tohoku earthquake and tsunami that destroyed the nuclear power plant of Fukushima, Japan, on March 11, 2011. A disaster that—in addition to the immediate destructive effects of the tsunami itself (18,500 dead or missing, 300,000 evacuated, some 125,000 buildings totally destroyed)—[105] resulted in a major drop in the electricity supply and caused economic stagnation in Japan.

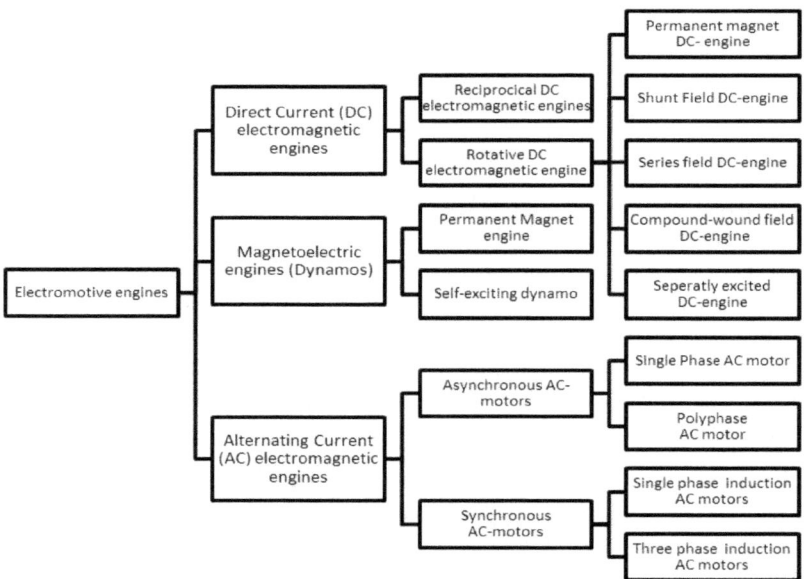

Figure 154: Overview of dominant electromotive engines

---

[105] Source: Damage Situation and Police Countermeasures. National Policy Agency of Japan, Emergy Disaster Countermeasures Headquarters.
http://www.npa.go.jp/archive/keibi/biki/index_e.htm (Accessed December 2014)

Dominant electromotive engines

This case study looked at the development of electromotive engines, from the first sparks of ideas to the engines that were implemented in a myriad of applications (Figure 154), a development that was at first obstructed by the limitations of battery power. This barrier was overcome by taking advantage of the reciprocity of the electric machine; it could also create electricity from rotative motion. The following abundance of electricity found its application in the fields of lighting and power. Then two other basic innovations, the arc light and the incandescent lamp, fueled its development.

In each of the clusters of innovations resulting in the direct current electic motor, the electric dynamo, and the alternating current induction motor, the dominant engines emerged. Engines that, due to their specific design, became succesful in their application. And that set the standard for the engiens developed later on.

Clusters of Businesses

Three basic innovations can be identified as being essential to the business development that took place over a century. It was the basic inventions of the DC motor, the dynamo, and the AC motor that became the nuclei of the progression of electricity into society. These basic innovations grew from the work of many individuals. Some were contributing to conceptual and theoretical insight, others—being more of a practical nature—contributed with their engineering skills, transferring the concept into working artifacts: the contributing innovations.

That process took time, as there was not one definitive moment nor one single person who had a magical "eureka" moment and created the invention. The contributors to the described development were the "gentlemen of science," more-or-less-privileged persons curious about the "nature of lightning." Other contributors were the electricians, persons not always gifted with theoretical insight, but who had the endurance to overcome practical problems and were able to create working artifacts.

And...., there were the innovator-entrepreneurs who created the enterprises that manufactured the actual machines. Many inventors became entrepreneurs exploiting their inventive work—such as Thomas Edison, who moved from a maverick trying to get incandescent lighting accepted, to a staunch opponent of the "dangerous" innovation of alternating current. He became involved in entrepreneurial activities, but his heart and soul were in inventing. This was similar to James Watt's situation (who invented the steam engine) when he wrote to his partner, James Boulton: "On the whole I find it is now full time to cease attempting to invent new things, or to attempt anything which is attended with any risk of not succeeding...Let us go on executing the things we understand, and leave the rest to younger men, who have neither money nor character to lose" (Scherer, 1965, p. 174). Edison was in the same position when he was more or less forced out during the creation of Edison General Electric. On February 8, 1890, he wrote to Villard to not oppose his "retirement from the lighting business, which will enable me to enter into fresh and congenial fields of work" (Paul A David, 1991, pp. 95-96).

The totality of all these efforts (Figure 154)—except perhaps for the early DC motor—resulted in several bonanzas of entrepreneurial activity: each basic innovation its own, different in its actual form, but nevertheless characterized by entrepreneurial activity, new startups, fierce competition, mergers, and acquisitions. The electromotive engines were applied in such a broad range of applications that the manufacturing of electrical products became a major manufacturing industry—an industry complemented by utility companies supplying and distributing the electricity needed.

Assuming one could distinguish the total development into three phases, we could group the developments in modern conceptual thinking in the following ways: research phase, development phase, and innovation phase (Table 22). A *grosso modo* indication is given of the different phases between the initial idea and the final saleable product. As phases and artifacts do overlap, this table shows only that the moment between the conception of the early principles and the implementation into grown-up applications is considerable.

Table 22: Development in phases for major innovations in electricity

	Research phase	Development phase		Innovation phase
	Early principle	*Early prototype*	*Early products*	*Saleable Product*
DC electric motor	1810–1830	1830–1835	1835–1855	1855–1880
Dynamo DC/AC	1830–1850	1850–1860	1860–1865	1860–1870
AC induction motor	1825–1835	1880–1885	1885–1890	1890–1895

Basic innovations: patents and their impact

Each of the basic innovations had its own *contributing innovations* and resulted in the *incremental innovations* that followed it. Patents often protected innovation. Some patents were for innovations that did not have an impact; other patents resulted in frantic infringement cases and patent wars. The issue of who had "priority" was not only a matter of honor, but also had its pecuniary consequences. Looking at the totality of the nineteenth century,

Table 23 presents those patents that were essential (directly or indirectly) in the development of the electromotive engine: the DC motor, the dynamo-electric generator (the dynamo), and the AC induction motor.

Table 23: Patents for basic innovations in the electromotive engines

Patent №.	Year	Patentee	Invention
US 132	1837	Davenport	DC motor
US 295.454	1888	Sprague	DC motor (railway applications)
US 494.978	1892	Crocker/Wheeler	DC motor (machine applications)
GB 806	1855	S. Hjorth	Dynamo-electric generator
GB 3.394	1886	S. A. Varley	Dynamo-electric generator
GB 261	1867	W. Siemens	Dynamo-electric generator
US 292.079	1884	Jonas Wenström	Dynamo-electric machine
US 381.968	1888	Nicola Tesla	Two-phase induction motor
US 390.439	1888	Charles Bradley	Two-phase induction motor
US 427.978	1890	Mikhail Dobrowolsky	Three-phase induction motor

To conclude

One can certainly conclude that society changed in the nineteenth century, and one of the major reasons was the new phenomenon called electricity.

> *Of the great construction projects of the last century, none has been more impressive in its technical, economic, and scientific aspects, none has been more influential in its social effects, and none has engaged more thoroughly our constructive instincts and capabilities than the electric power system…Electric power systems embody the physical, intellectual, and symbolic resources of the society that constructs them…In a sense electric power systems, like so much other technology, are both causes and effects of social change* (Hughes, 1993, pp. 1-2).

References

Abernathy, W. J., & Clark, K. B. (1985). Innovation: Mapping the winds of creative destruction. *Research Policy, 14*(1), 3-22. doi: http://dx.doi.org/10.1016/0048-7333(85)90021-6

Adams, M. (2012). Born to Invent *Lee de Forest* (pp. 1-39): Springer New York.

Ames Hydroelectric Generating Plant. (1891). *Milestones*. Retrieved August 15th, 2013, from http://www.ieeeghn.org/wiki/index.php/Milestones:Ames_Hydroelectric_Generating_Plant,_1891

Ampère, A.-M. (1821). *Mémoire sur l'action mutuelle de deux courants électriques, sur celle qui existe entre un courant électrique et un aimant ou le globe terrestre, et celle de deux aimants l'un sur l'autre*.

Ampère, A.-M. (1826). *Description d'un appareil électro-dynamique*: Bachelier.

Ampère, A. M. (1826). *Théorie des phénomènes électro-dynamiques: uniquement déduite de l'expérience*: Méquignon-Marvis.

Arapostathis, S. (2013). 7. Contested inventors: British patent disputes and the culture of invention in the late nineteenth century. *Knowledge Management and Intellectual Property*, 145.

Arapostathis, S., & Gooday, G. (2013). *Patently contestable: Electrical technologies and inventor identities on trial in Britain*: The MIT Press.

Asztalos, P. (1986). Centenary of the Transformer. *Periodica Polytechnica, 30*(1).

Babbage, C., & Herschel, J. F. W. (1825). Account of the Repetition of M. Arago's Experiments on the Magnetism Manifested by Various Substances during the Act of Rotation. *Philosophical Transactions of the Royal Society of London, 115*, 467-496.

Baily, W. (1879). A Mode of producing Arago's Rotation. *Proceedings of the Physical Society of London, 3*(1), 115.

Barreyre, N. (2011). The Politics of Economic Crises: The Panic of 1873, the End of Reconstruction, and the Realignment of American Politics. *The Journal of the Gilded Age and Progressive Era, 10*(04), 403-423. doi: doi:10.1017/S1537781411000260

Beauchamp, K. G. (1997). *Exhibiting Electricty* (Vol. 21): Iet.

Bergek, A., Jacobsson, S., Carlsson, B., Lindmark, S., & Rickne, A. (2008).

Analyzing the functional dynamics of technological innovation systems: A scheme of analysis. *Research Policy, 37*(3), 407-429. doi: http://dx.doi.org/10.1016/j.respol.2007.12.003

Berger, H., & Spoerer, M. (2001). Economic Crises and the European Revolutions of 1848. *The Journal of Economic History, 61*(2), 293-326. doi: 10.2307/2698022

Bernardi, W. (2000a). The controversy on animal electricity in eighteenth-century Italy: Galvani, Volta and others. *Nuova Voltianas, Bevilacqua and Fregonese, eds.(Milan: Hoepli)*, 101-114.

Bernardi, W. (2000b). The controversy on animal electricity in eighteenth-century Italy: Galvani, Volta and others. *Nova Voltiana: Studies on Volta and His Times, 1*, 113.

Blalock, T. J. (2011). Ampere, New Jersey [History]. *Power and Energy Magazine, IEEE, 9*(3), 78-91. doi: 10.1109/MPE.2011.940407

Blondel, C. (1997). Electrical instruments in 19th century France, between makers and users. *History and Technology, an International Journal, 13*(3), 157-182.

Blondel, C. W., B. (2012). In Search of a Newtonian Law of Electrodynamics (1820-1826). http://www.ampere.cnrs.fr/parcourspedagogique/zoom/courant/formule/index-en.php

Bowers, B. (1972). The eccentric electromagnetic engine—a chapter from the very early history of the electric motor. *Electronics & Power, 18*(7), 269-272.

Bowers, B. (1975). Charles wheatstone—experimental philosopher. *Contemporary Physics, 16*(5), 499-512.

Bowers, B. (2001). *Sir Charles Wheatstone FRS*.

Bowers, B., & Symons, L. (2006). *Curiosity perfectly satisfied : Faraday's travels in Europe, 1813-1815*. London: Institution of Engineering and Technology in association with the Science Museum.

Brandon, C. (2009). *The electric chair: An unnatural American history*: McFarland.

Brittain, J. C. (1995). Benjamin G. Lamme and Giant Generators. *Proceedings of the IEEE, 83*(11), 1593-1594.

Brittain, J. C. (2002). Charles F. Scott: A pioneer in electrical power engineering. *Industry Applications Magazine, IEEE, 8*(6), 6-8.

Brittain, J. E. (1974). The international diffusion of electrical power technology, 1870-1920. *The Journal of Economic History, 34*(1), 108-121.

Byrn, E. W. (1900). *The progress of invention in the nineteenth century*: Munn & Company.

Caneva, K. L. (1978). From Galvanism to electrodynamics: The transformation of German physics and its social context. *Historical studies in the physical sciences*, 63-159.

Cantor, G. (1989). Why Was Faraday Excluded from the Sandemanians in 1844? *The British Journal for the History of Science, 22*(4), 433-437. doi: 10.2307/4026920

Capolino, G. A. (2004). André Blondel - French Scientist and Engineer. *IEEE Industry Applications Magazine*(May/June).

Cardwell, D. (1976). Science and Technology: The Work of James Prescott Joule. *Technology and Culture, 17*(4), 674-687. doi: 10.2307/3103674

Cardwell, D. (1992). On Michael Faraday, Henry Wilde, and the dynamo. *Annals of Science, 49*(5), 479-487.

Carlson, W. B. (1995). The Coordination of Business Organization and Technological Innovation within the Firm: A Case Study of the Thomson-Houston Electric Company in the 1880s. In E. Naomi R. Lamoreaux; Daniel M.G. Raff (Ed.), *Coordination and Information: Historical Perspectives on the*

Organization of Enterprise (pp. 55-100): University of Chicago Press.

Carlson, W. B. (2013). *Tesla: Inventor of the Electrical Age*: Princeton University Press.

Cases argued and determined in the Circuit Courts of Appeals and Circuit and District Courts of the United States. . (1904). *The Federal Reporter, 129*(June-July), 215-201.

Ciok, Z. K., L. Nowakowski, R. Szymczak, P. . (2009). Michał Doliwo-Dobrowolski - współtwórca cywilizacji technicznej XX wieku. *Wiadomości Elektrotechniczne, R. 77*(Nr. 1), 38-49.

Currier, D. P. (1857). A Biographical History of induction coils: Essay by Dean P. Currier (dpcurr@ aol. com) posted on www. radiantslab. com/quackmed/Deanbio. html [8] Rev N. Callan, On the Induction Apparatus, Philosophical Magazine.

Dalzell, F. (2010). *Engineering invention*: The MIT Press.

David, P. A. (1991). The hero and the herd in technological history: Reflections on Thomas Edison and the battle of the systems. *Favorites of fortune*, 72-119.

David, P. A., & Bunn, J. A. (1988). The economics of gateway technologies and network evolution: Lessons from electricity supply history. *Information Economics and Policy, 3*(2), 165-202. doi: http://dx.doi.org/10.1016/0167-6245(88)90024-8

Davy, H. (1821). On the Magnetic Phenomena Produced by Electricity. *Philosophical Transactions of the Royal Society of London, 111*, 7-19. doi: 10.2307/107599

Davy, J. (1839). *The Collected Works of Sir Humphry Davy* (Vol. V). London: Smith, Elder & Co.

Day, L. M., I. (2013). Kapp, Gisbert Johann Eduard Karl *Biographical Dictionary of the History of Technology* (pp. 864): Routledge.

Devezas, T. C. (2005). Evolutionary theory of technological change: State-of-the-art and new approaches. *Technological Forecasting and Social Change, 72*(9), 1137-1152. doi: http://dx.doi.org/10.1016/j.techfore.2004.10.006

Devine, W. D. (1983). From shafts to wires: Historical perspective on electrification. *Journal of Economic History, 43*(02), 347-372.

Dibner, B. (1964). James Clerk Maxwell. *Spectrum, IEEE, 1*(12), 50-56. doi: 10.1109/MSPEC.1964.6501276

Dood, K. J., Leland, I. A., & Kline, R. R. (1989). Tesla and the Induction Motor. *Technology and Culture, 30*(4), 1013-1023. doi: 10.2307/3106202

Doppelbauer, M. (2012). The invention of the electric motor 1800-1854. Retrieved June 29th, 2013, from http://www.eti.kit.edu/english/1376.php

Douglass, J. N. (1879). The Electric Light Applied to Lighthouse Illumination. *Proceedings of the Institution of Civil Engineers, 108*, 77-111.

Dredge, J. (1882). *Electric Illumination* (Vol. I). New York: John Wiley and Sons.

du Fay, C. (1734). Two Kinds of Electrical Fluid: Vitreous and Resinous. *Philosophical Transactions of the Royal Society of London, 38*.

Duijn, J. J. v. (1983). *The long wave in economic life*: Allen & Unwin London.

Dyer, F. L., & Martin, T. C. (1910). *Edison, His Life And Inventions*. New

York: Harper & Brothers Publishers

Early Electrification of Buffalo. (2004). Retrieved September 12th 2013, from http://www.ieeeghn.org/wiki/index.php/Early_Electrification_of_Buffalo:_The_Beginning_of_Central_Station_Service

Edison Electric Light, C. (1887). *A warning from the Edison Electric Light Co.* [New York, N.Y.]: The Company.

Edison, T. A. (2013). The Thomas Edison Papers. Retrieved August 9th 2013, from National Archives, Washington, D.C. http://edison.rutgers.edu/NamesSearch/glocpage.php3?gloc=W100DD

Evans, D. M. (1849). *The commercial crisis, 1847-1848*. London: Letts, Son and Steer.

Evans, R. J. W. (1988). *Epidemics and Revolutions: Cholera in Nineteenth-Century Europe*: Oxford University Press on behalf of The Past and Present Society.

Evans, R. J. W. (2000). Liberalism, Nationalism, and the Coming of the Revolution. *The Revolutions in Europe 1848–1849: From Reform to Reaction, 9-26*.

Faraday, M. (1821). Historical sketch of electro-magnetism. *Annals of Philosophy, 2*, 195-200.

Faraday, M. (1832). Experimental researches in electricity. *Philosophical transactions of the Royal Society of London, 122*, 125-162.

Faraday, M. (1835). XXXVII. Reply to Dr. John Davy's "Remarks on certain statements of Mr. Faraday contained in his 'researches in electricity'" To Richard Phillips, Esq.

Faraday, M., & James, F. A. (1991). *The Correspondence of Michael Faraday: 1811-December 1831, letters 1-524*.

Feldenkirchen, W. (1994). *Werner von Siemens: inventor and international entrepreneur*: Ohio State University Press.

Foran, J. The day they turned the falls on: the invention of the universal electrical power system. *Internet article, Case studies in Science, http://ublib. buffalo. edu/ libraries/ projects/ cases/ niagra. htm*.

Franklin, B. (1751). A Letter of Benjamin Franklin, Esq; to Mr. Peter Collinson, F. R. S. concerning an Electrical Kite. *Philosophical Transactions (1683-1775), 47*, 565-567. doi: 10.2307/105108

Gardiner, K. R., & Gardiner, D. L. (1965). André-Marie Ampère and His English Acquaintances. *The British Journal for the History of Science, 2*(3), 235-245. doi: 10.2307/4024937

Geddes, L. A., & Hoff, H. E. (1971). The discovery of bioelectricity and current electricity The Galvani-Volta controversy. *Spectrum, IEEE, 8*(12), 38-46.

Goldstone, J. A. (2011). Understanding the revolutions of 2011: weakness and resilience in Middle Eastern autocracies. *Foreign Aff., 90*, 8.

Gooding, D. (1985). Experiment and concept formation in electromagnetic science and technology in England in the 1820s. *History and Technology, 2*(2), 151-176. doi: 10.1080/07341518508581638

Gramme Electrical Co. V. Arnoux & Hochhausen Electric Co. (1883).

Griffith, W. P. (1983). Priestley in London. *Notes and Records of the Royal Society of London, 38*(1), 1-16. doi: 10.2307/531343

Grönberg, P.-O. (2003). *Learning and Returning: Return Migration of Swedish Engineers from the United*

States, 1880-1940: Department of Historical Studies, Umeå University, Umeå, Sweden.

Haldane Gee, W. W. (1920). Henry Wilde *Memoirs and Proceedings of the Manchester Literary and Philosophical Society*: Manchester : The Society.

Hammond, J. W. (1941). *Men and volts: the story of general electric*: JB Lippincott Company.

Heilbron, J. L. (1979). *Electricity in the 17th and 18th Century: A Study of Early Modern Physics*: Univ of California Press.

Henry, J. (1832). On the production of currents and sparks of electricity from magnetism. *American Journal of Science and Arts, XXII*(July).

Henry, J. (1839). Contributions to Electricity and Magnetism, number 3: On Electrodynamic Induction'. *American Philosophical Society Transactions, 6*, 303-338.

Higgs, R. W. H., & Brittle, J. R. (1878). Some recent improvements in dynamic-electric apparatus. *Minutes of Proceedings, Vol.52*(Issue 1878), 57-98. doi: 10.1680/imotp.1878.22459

Hobsbawm, E. J. (1952). The Machine Breakers. *Past & Present*(1), 57-70. doi: 10.2307/649989

Hobsbawm, E. J. (2010). *Age of Revolution 1789-1848*: Weidenfeld & Nicolson.

Hofmann, J. R. (1995). *André-Marie Ampère: Enlightenment and Electrodynamics* (Vol. 7): Cambridge University Press.

Hopkinson, J., & Hopkinson, E. (1886). Dynamo-Electric Machinery. *Philosophical Transactions of the Royal Society of London, 177*, 331-358.

Hughes, T. P. (1962). British Electrical Industry Lag: 1882-1888. *Technology and Culture, 3*(1), 27-44. doi: 10.2307/3100799

Hughes, T. P. (1993). *Networks of power: electrification in Western society, 1880-1930*: Johns Hopkins University Press.

Hunt, L. (1973). The early history of gold plating. *Gold Bulletin, 6*(1), 16-27.

Imerito, T. (2013). The Westinghouse Legacy. from http://www.pghtech.org/news-and-publications/teq/article.aspx?Article=1759

Jeffery, J. V. (1997). The Varley family: engineers and artists. *Notes and Records of the Royal Society of London, 51*(2), 263-279.

Jonas Wenström (1928). *Teknisk Tidskrift*, 454-458.

Jones, B., & Faraday, M. (2010). *The life and letters of Faraday* (Vol. 2): Cambridge University Press.

Khantine-Langlois, F. (2005). Lenz, un savant méconnu. *Bulletin de l'Union des physiciens*(876), 705-712.

King, W. J. (1962). *The Development of Electrical Technology in the 19th Century.* Washington Governement Pr.: Smithsonian Institution.

Kipnis, N. (1987). Luigi Galvani and the debate on animal electricity, 1791–1800. *Annals of Science, 44*(2), 107-142. doi: 10.1080/00033798700200151

Kipnis, N. (2005). Chance in Science: The Discovery of Electromagnetism by H.C. Oersted. *Science & Education, 14*(1), 1-28. doi: 10.1007/s11191-004-3286-0

Klein, M. (2010). *The power makers: Steam, electricity, and the men who invented modern America*: Bloomsbury Publishing USA.

Kline, R. (1987). Electricity and Socialism: The Career of Charles P. Steinmetz. *Technology and Society Magazine, IEEE, 6*(2), 9-17. doi: 10.1109/MTAS.1987.5010093

Kline, R. (1987). Science and Engineering Theory in the Invention and Development of the Induction Motor, 1880-1900. *Technology and Culture, 28*(2), 283-313. doi: 10.2307/3105568

Klotz, I. M. (1993). "Misconduct" in science: Quis custodiet ipsos custodes. *Academic Questions, 6*(4), 37-48.

Knight, D. M. (2000). Humphry Davy: science and social mobility. *Endeavour, 24*(4), 165-169. doi: http://dx.doi.org/10.1016/S0160-9327(00)01326-0

Kurzweil, P. (2010). Gaston Planté and his invention of the lead–acid battery—The genesis of the first practical rechargeable battery. *Journal of Power Sources, 195*(14), 4424-4434. doi: http://dx.doi.org/10.1016/j.jpowsour.2009.12.126

Larmor, J. (1937). *Origins of Clerk Maxwell's Electric Ideas as Described in Familiar Letters to William Thomson*: Cambridge.

Lee, L.-C. A. G. (1932). The Varley brothers: Cromwell Fleetwood Varley and Samuel Alfred Varley. *Electrical Engineers, Journal of the Institution of, 71*(432), 958-964. doi: 10.1049/jiee-1.1932.0174

Leitch, A., & Leitch, A. (1978). *A Princeton companion*: Princeton University Press Princeton, PA.

Lenz, E. (1834). Über die Bestimmung der Richtung der durch elektrodynamische Vertheilung erregten galvanischen Ströme. [Source: Gallica.bnf.fr / Bibliothèque nationale de France]. *Annalen der Physik, 107*(31), 483-494.

Leupp, F. E. (1918). *George Westinghouse: his life and achievements*: Little, Brown, and company.

Lipsey, R. G., Carlaw, K. I., & Bekar, C. T. (2005). *Economic Transformations: General Purpose Technologies and Long-Term Economic Growth*: Oxford University Press.

MacKechnie Jarvis, C. (1955a). The history of electrical engineering. Part 4: Machinery for the new light: part 2. *Electrical Engineers, Journal of the Institution of, 1*(9), 566-574. doi: 10.1049/jiee-3.1955.0210

MacKechnie Jarvis, C. (1955b). The history of electrical engineering. Part 4: Machinery for the new light: Part 2. *Journal of the IEE, 1*(9), 566-574.

MacLeod, C. (2012). Reluctant Entrepreneurs: Patents and State Patronage in New Technosciences, circa 1870–1930. *Isis, 103*(2), 328-339. doi: 10.1086/666359

Martin, T. C., & Coles, S. L. (1919). *The story of electricity* (Vol. 2): Story of electricity Company, MM Marcy.

Maxwell, J. C. (1861). On physical lines of force: Part I.–The Theory of Molecular Vortices applied to Magnetic Phenomena. *The London, Edinburgh, and Dublin Philosophical Magazine and Journal of Science, 21*(139), 161-175.

Maxwell, J. C. (1864). On Faraday's lines of force. *Transactions of the Cambridge Philosophical Society, 10*, 27.

Maxwell, J. C. (1865). A dynamical theory of the electromagnetic field. *Philosophical Transactions of the Royal Society of London, 155*, 459-512.

Maxwell, J. C. (1873). A treatise on electricity and magnetism. *Clarendon Press series*.

Maxwell, J. C. (1990). *The scientific letters and papers of James Clerk Maxwell: 1862-1873. Vol. 2* (Vol. 1): CUP Archive.

McPartland, D. S. (2006). *Almost Edison: How William Sawyer and Others Lost the Race to Electrification*: ProQuest.

Mensch, G. (1979). *Stalemate in technology: innovations overcome the depression*: Ballinger Cambridge, Mass.

Michalowicz, J. C. (1948). Origin of the electric motor. *Electrical Engineering, 67*(11), 1035-1040.

Mokyr, J. (1980). Industrialization and Poverty in Ireland and the Netherlands. *The Journal of Interdisciplinary History, 10*(3), 429-458. doi: 10.2307/203187

Moran, R. (2007). *Executioner's current: Thomas Edison, George Westinghouse, and the invention of the electric chair*: Random House Digital, Inc.

Morris, E. (2007). From horse power to horsepower. *ACCESS Magazine, 1*(30).

Morton Jr, D. L. (2002). Reviewing the history of electric power and electrification. *Endeavour, 26*(2), 60-63. doi: http://dx.doi.org/10.1016/S0160-9327(02)01422-9

Neidhofer, G. (2007). Early three-phase power. *Power and Energy Magazine, IEEE, 5*(5), 88-100. doi: 10.1109/MPE.2007.904752

Nishimura, S. (2012). The rise of the patent department: A case study of Westinghouse Electric and Manufacturing Company.

Nye, D. E. (1990). *Electrifying America: social meanings of a new technology, 1880-1940*. Cambridge Massachusetts: The MIT Press.

Otto Sibum, H. (2003). Experimentalists in the Republic of Letters. *Science in Context, 16*(1-2), 89-120. doi: doi:10.1017/S0269889703000747

Passer, H. C. (1972). *The electrical manufacturers, 1875-1900: A study in competition, entrepreneurship, technical change, and economic growth*: Arno Press.

Piccolino, M. (1998). Animal electricity and the birth of electrophysiology: the legacy of Luigi Galvani. *Brain research bulletin, 46*(5), 381-407.

Post, R. C. (1972). The Page Locomotive: Federal Sponsorship of Invention in Mid-19th-Century America. *Technology and Culture, 13*(2), 140-169.

Post, R. C. (1976). Stray sparks from the induction coil: The volta prize and the page patent. *Proceedings of the IEEE, 64*(9), 1279-1286.

Priestley, J. (1769a). *A familiar introduction to the study of electricity*.

Priestley, J. (1769b). *The history and present state of electricity: with original experiments*: J. Dodsley in Pall-Mall, J. Johnson and J. Payne in Paternoster row, and T. Cadell (successor to Mr. Millar) in the Strand.

Priestley, J. (1771). *An Essay on the First Principles of Government: and on the nature of political, civil, and religious liberty*: J. Johnson.

Priestley, J. (1782). Institutes of Natural and Revealed Religion, 2 vols., (Birmingham, 1782). *The Theological & Miscellaneous Works of Joseph Priestley, LL. DFRS etc., ed. JT Rutt, 25*, 1817-1835.

Priestley, J. (1786). *Experiments and observations relating to various branches of natural philosophy with a continuation of the observations on air* (Vol. 3): printed by Pearson and Rollason for J. Johnson... London.

Priestley, J., & DFRS, L. (1775). *Experiments and observations on different kinds of air*.

Priestley, J., & Johnson, J. (1792). *An Appeal to the Public, on the Subject of the Riots in Birmingham, Part II: To which is Added, a Letter from W. Russell, Esq. to the Author*: J. Johnson, St. Paul's Church Yard.

Prout, H. G. (1921). *A life of George Westinghouse*: The American society of mechanical engineers.

Reeves, J. (1887). *The Rothschilds: the financial rulers of nations*: Sampson Low, Marston, Searle, and Rivington.

Reynolds, T. S., & Bernstein, T. (1989). Edison and'the chair'. *Technology and Society Magazine, IEEE, 8*(1), 19-28.

Rivers, I., & Wykes, D. L. (2008). *Joseph Priestley, Scientist, Philosopher and Theologian*: Oxford University Press.

Robertson, P. S. (1952). *Revolutions of 1848: a social history* (Vol. 1025): Princeton University Press.

Rose, R. B. (1960). The Priestley riots of 1791. *Past & Present*(18), 68-88.

Ross, S. (1965). The Search for Electromagnetic Induction: 1820-1831. *Notes and Records of the Royal Society of London, 20*(2), 184-219. doi: 10.2307/3519873

Rushmore, D. B. (1905). The alternating current generator. *Journal of the Franklin Institute, 160*(4), 253-274. doi: http://dx.doi.org/10.1016/S0016-0032(05)90146-4

Scherer, F. M. (1965). Invention and Innovation in the Watt-Boulton Steam-Engine Venture. *Technology and Culture, 6*(2), 165-187. doi: 10.2307/3101072

Schiffer, M. B. (2002). Studying Technological Differentiation: The Case of 18th-Century Electrical Technology. *American Anthropologist, 104*(4), 1148-1161. doi: 10.2307/3567103

Schiffer, M. B. (2005). The electric lighthouse in the nineteenth century: Aid to navigation and political technology. *Technology and Culture, 46*(2), 275-305.

Schiffer, M. B. (2006). *Draw the lightning down: Benjamin Franklin and electrical technology in the age of enlightenment*: Univ of California Press.

Schonland, B. F. J. (1952). The work of Benjamin Franklin on thunderstorms and the development of the lightning rod. *Journal of the Franklin Institute, 253*(5), 375-392. doi: http://dx.doi.org/10.1016/0016-0032(52)90717-5

Schonland, B. F. J. (1952). The work of Benjamin Franklin on thunderstorms and the development of the lightning rod. *Journal of Franklin Institute, 253*(5), 375-392.

Schumpeter, J. A. (1939). *Business cycles; a theoretical, historical, and statistical analysis of the capitalist process (Fels)* (1st ed.). New York, London,: McGraw-Hill Book Company, inc.

Schumpeter, J. A., & Opie, R. (1934). *The theory of economic development; an inquiry into profits, capital, credit, interest, and the business cycle.* Cambridge, Mass.,: Harvard University Press.

Shaver, L. (2012). Illuminating Innovation: From Patent Racing to Patent War. *Wash. & Lee L. Rev., 69*, 1891.

Skrabec, Q. R. (2007). *George Westinghouse: Gentle Genius*: Algora Publishing.

Smil, V. (2005). *Creating the Twentieth Century: technical innovations of 1867-1914 and their lasting impact*: Oxford University Press, USA.

Smith, S. (1912). *Søren Hjorth: inventor of the dynamo-electric principle*: Printed by J. Jørgensen & co.(MA Hannover).

Stanley, W. (1912). Alternating-current development in America. *Journal of the Franklin Institute, 173*(6), 561-580. doi: http://dx.doi.org/10.1016/S0016-0032(12)90176-3

Steinle, F. (2002). Experiments in history and philosophy of science. *Perspectives on Science, 10*(4), 408-432.

Taylor, M. (2000). The 1848 revolutions and the British empire. *Past & Present*(166), 146-180.

Terry, C. A. (1898). Oliver Blackburn Shallenberger: A memorial. *American Institute of Electrical Engineers, Transactions of the, 15*(1), 744-753.

Tesla, N. (2007). *My inventions: the autobiography of Nikola Tesla*: Wilder Publications.

Thompson, S. P. (1896). *Dynamo-electric machinery: a manual for students of electrotechnics*: E. & FN Spon.

Thompson, S. P. (1898). *Michael Faraday: his life and work*: The Macmillan.

Thomson, W. (1881). ON THE SOURCES OF ENERGY IN NATURE AVAILABLE TO MAN FOR THE PRODUCTION OF MECHANICAL EFFECT. *Science, os-2*(68), 475-478. doi: 10.1126/science.os-2.68.475

Usher, A. P. (1929). *A history of mechanical inventions*. New York: McGraw-Hill Book Company.

Van Trump, J. D. (1959). "Solitude" and The Nether Depths: The Pittsburgh Estate of George Westinghouse and its Gas Well. *Western Pennsylvania History, 42*(2), 155-172.

Vanhaute, E., O'Grada, C., & Paping, R. (2007). The European subsistence crisis of 1845-1850. A comparative perspective.

Volta, A. (1800). On the Electricity Excited by the Mere Contact of Conducting Substances of Different Kinds. In a Letter from Mr. Alexander Volta, F. R. S. Professor of Natural Philosophy in the University of Pavia, to the Rt. Hon. Sir Joseph Banks, Bart. K. B. P. R. S. *Philosophical Transactions of the Royal Society of London, 90*, 403-431. doi: 10.2307/107060

Ward-Perkins, C. N. (1950). The commercial crisis of 1847. *Oxford Economic Papers, 2*(1), 75-94.

White, J. H. (2005). War of the Wires: A Curious Chapter in Street Railway

History. *Technology and Culture, 46*(2), 374-384. doi: 10.2307/40060853

Wilde, H. (1867). Experimental Researches in Magnetism and Electricity. *Philosophical Transactions of the Royal Society of London, 157*, 89-107. doi: 10.2307/108969

Wilde, H. (1900). *Correspondence in the matter of the Society of Arts and Henry Wilde, D. Sc., FRS: on the award to him of the Albert Medal, 1900: and on the invention of the dyamo-electric machine*: H. Rawson & Co., Printers.

Williams, L. P. (1965). *Michael Faraday: a biography*: Chapman and Hall London.

Williams, L. P. (1983). What Were Ampere's Earliest Discoveries in Electrodynamics? *Isis, 74*(4), 492-508. doi: 10.2307/232209

Wilson, C. H. (1939). The Economic Decline of the Netherlands. *The Economic History Review, 9*(2), 111-127. doi: 10.2307/2590218

Wisniak, J. (2004). André-Marie Ampère. The chemical side. *Educación química, 15*, 2.

Zadoks, J. (2008). The potato murrain on the European continent and the revolutions of 1848. *Potato Research, 51*(1), 5-45.

About the author

Drs.Ir.Ing. B. J. G. van der Kooij (1947) obtained his MBA in 1975 (thesis: Innovation in SMEs) at the *Interfaculteit Bedrijfkunde* (now part of the Rotterdam Erasmus University), followed by obtaining his MSEE (thesis: Micro-electronics) at the Delft University of Technology in 1977.

He started his career as assistant to the board of directors of Holec NV, a manufacturer of electrical power systems that employed about eight thousand people at that time. His responsibilities were in the field of corporate strategy and innovation of Holec's electronic activities. Travelling extensively to Japan and California, he became well known as a Dutch guru on the topic of innovation and microelectronics.

From 1982 to 1986, he was a member of the Dutch Parliament (*Tweede Kamer der Staten Generaal*) and spokesman on the fields of economic, industrial, science, innovation, and aviation policy. He became known as the first member to introduce the personal computer in Parliament, but his work on topics like the TNO-Act, Patent Act, Chips-Act, and others went largely unnoticed.

After the 1986 elections and the massive loss for his party (VVD), he retired from politics and became a part-time professor (*Buitengewoon Hoogleraar*) at the Eindhoven University of Technology. His field was the management of innovation. He also started his own company, Ashmore Software BV, in 1986, where he developed software for professional tax applications on personal computers.

After finishing these activities in 2003, he became a real-estate project developer and, in 2009, a real-estate consultant till his retirement in 2013. Since innovation was the focus of all his corporate, entrepreneurial, political, and scientific life, he wrote three books on the subject and published several articles. In his first book, he explored the technological dimension of innovation (the pervasive role of microelectronics). His second book focused on the management of innovation and the human role in the innovation process. And in his third book, he formulated "Laws of Innovation" based on the Dutch societal environment in the 1980s.

In 2012 he started studying the topic of innovation again. In 2013 Prof. Dr. Cees van Beers accepted him as a PhD candidate at the TU-Delft. His focus is on the theory of innovation, and his aim is to develop a multidimensional model explaining innovation. To accomplish this aim, he creates extensive and detailed case studies observing the inventions of the steam engine, the electromotive engines, the communicating engines, and the computing engines. He studies their characteristics from a multidisciplinary perspective (economic, technical, social).

Van de Kooij is married and spends a great deal of his time working in the south of France.

www.ingramcontent.com/pod-product-compliance
Lightning Source LLC
Chambersburg PA
CBHW051800170526
45167CB00005B/1823